Methods and Procedures for Building Sustainable Farming Systems

Ana Alexandra Marta-Costa • Emiliana Silva
Editors

Methods and Procedures for Building Sustainable Farming Systems

Application in the European Context

Springer

Editors
Dr. Ana Alexandra Marta-Costa
Centre for Transdisciplinary Development Studies
University of Trás-os-Montes and Alto Douro
Quinta de Prados, Apartado 1013
5001-801 Vila Real
Portugal

Dr. Emiliana Silva
Centre of Applied Economics Studies of the Atlantic
Agricultural Sciences Department
University of Azores
Rua Capitão João d'Ávila
9700-042 Angra do Heroísmo
Açores, Portugal

ISBN 978-94-007-5002-9 ISBN 978-94-007-5003-6 (eBook)
DOI 10.1007/978-94-007-5003-6
Springer Dordrecht Heidelberg New York London

Library of Congress Control Number: 2012950001

Kind permission was given by Elsevier Limited, Oxford, UK for the use of chapters 4 and 6.
Chapter 4 was originally published in 2010 from "Environmental Impact Assessment Review, 30, 71–81. Binder C.R., Feola G., Steinberger J., Considering the normative, systemic and procedural dimensions in indicator-based sustainability assessments in agriculture."
Chapter 6 was originally published in 2012 from "Environmental Impact Assessment Review, 32, 170–180. Van Passel, S. & Meul, M., Multilevel and multi-user sustainability assessment of farming systems."

Printed on acid-free paper

Springer is part of Springer Science+Business Media (www.springer.com)

Foreword

As I was then president of the European Group of the International Farming Systems Association, it was an honour for me to host the 9th European IFSA Symposium in Vienna in July 2010. There were, as is always the case in IFSA Symposia, many workshops. But of course, some workshop convenors stand out as being more active than others, not least due to their commitment to make the results of the discussions widely available. The convenors of the workshop on 'Methods and Procedures for Sustainable Farming Systems' were definitely such a group, and I am delighted to see that Alexandra Marta-Costa and Emiliana Silva are now publishing the key papers of their workshop as a book. In the time since the IFSA Symposium, the papers have been revised and reworked several times so that they are now mature and well polished, ready for publication. This book is a wonderful testimony of the quality of the contributions and discussions in that workshop.

The 15 chapters which report the results of separate studies focus on very different issues, in different contexts and using different modelling approaches. All include environmental sustainability considerations, and most include economic aspects. For these two aspects of sustainability, assessment tools have been developed for over 20 years and now cover a very broad array of approaches suitable for very different contexts and issues. The chapters describe various approaches, including multi-criteria models, mathematical programming, indicator-based methods, tools based on hierarchical processes and tools structured along the life cycle of a product or focusing on the local interactions between farms and other activities in the territory. The approaches thus differ in the scale of analysis (farm level vs. territory or landscape level) and the types of interactions taken into account (e.g. use of agrochemicals, nutrient leaching, number of farm activities, crop growth, energy use, costs). Many models couple the ecological and the economic assessment since the cost of various measures (or of non-action) is often key considerations for policy makers. Similarly, the trade-off between ecological and economic targets is a key concern for farmers, as too often environmental protection measures are costly, and these costs tend to be borne by the farmer.

But the chapters do not only cover methods which assess – and thus can contribute to – the sustainable use of natural resources and the economic viability of farms, a few also specifically assess energy use. This may well be seen as a response to the pressure of agriculture to contribute to climate change mitigation, among others, by reducing their energy use. Researchers are thus called upon to develop novel models which make this aspect explicit.

Assessing social sustainability is not as well represented in the chapters, and the approaches are still exploratory. Indeed, in agricultural research, social sustainability is often equated with economic aspects, based on the assumption that assuring an adequate income for farm families is the one key concern. As a result, other considerations, such as quality of life or work satisfaction, are not prominent concerns taken up by researchers working on the sustainability of farming systems.

While a number of the approaches presented in this book do include some participatory components, many are still based on a top-down approach, which positions researchers as the experts and farmers – as well as other stakeholders – as the recipients of this expertise. This may be linked to the fact that many tools presented are based on quantitative models, which tend to dictate the type of information and relationships included and reduce the variables to those that can be quantified and where data are available. Also, the model structure of such formalised models can be difficult to communicate to stakeholders. Doubtlessly, these models provide valuable insights that are well grounded in scientific knowledge, but experience has shown that these insights are not always used by decision makers. Various authors have thus pointed out the advantages of participatory approaches in framing the research questions so that the results are relevant to the needs of decision makers and thereby increasing the likelihood that they will be used.

Participatory approaches also have the advantage that they allow to take into account different perceptions and views of the 'problem', thus highlighting the differences both in what goals should be achieved and what means to achieve these goals are seen as suitable. I am thus particularly delighted to see that several chapters in this book take on the challenge of further developing participatory, bottom-up approaches and hope they may inspire further work in this area. I would especially welcome to see bottom-up work at the farm level, thus effectively empowering farmers to better understand the interactions between ecological, economic and social aspects, as well as the trade-offs which a specific choice is likely to imply. Such participatory approaches are valuable contributions to social learning, raising the awareness for the interdependencies and thus strengthening systemic thinking.

In its methodological diversity, this is doubtlessly a timely book. The valuable compilation of different approaches will allow the reader to assess the strength and the weakness of each, become aware of the variety of tools that allow to address specific aspects of sustainability and thus inform the choice of an adequate approach for a new study. It is thus with a warm 'thank you' that I would like to congratulate Alexandra Marta-Costa and Emiliana Silva for their success in compiling this valuable book.

Department of Economic and Social Sciences
University of Natural Resources and Life Sciences,
Vienna (Austria)

Ika Darnhofer

Acknowledgements

This book could not be possible without the dedicated and interested contribution of persons and institutions. We begin by thanking the efforts of the authors to attend to our challenge. They agreed to contribute with the texts that are now published. These are based in submissions and their comments presented in the 2.1 Methods and Procedures for Sustainable Farming Systems workshop on the 9th European International Farming Systems Association (IFSA) Symposium, which took place in Vienna, Austria, in July of 2010. It was organised jointly by the Centre for Transdisciplinary Development Studies (CETRAD) of the University of Trás-os-Montes and Alto Douro and the Centre of Applied Economics Studies of the Atlantic (CEEAplA) of the University of Azores. We also express our appreciation to the Steering Committee of IFSA, especially to her president in 2010 – Ika Darnhofer – for the availability of the essential facilities and resources to a proper realisation of the event as well as all her support and encouragement to the publishing of this book. Finally, a kindly and generously acknowledgement should be done to the main referees (Alexandra Marta-Costa, Emiliana Silva and Miguel SottoMayor) that carried out all the process of reviewing of the writings that are now part of this book.

The editors

Contents

Contributors

Alexandra Sintori Department of Agricultural Economics and Rural Development, Agricultural University of Athens (AUA), Athens, Greece

Alvaro Rocca Department of Agricultural and Environmental Sciences, University of Udine, Udine, Italy

Ana Alexandra Marta-Costa Centre for Transdisciplinary Development Studies (CETRAD), University of Trás-os-Montes and Alto Douro (UTAD), Vila Real, Portugal

Andrea Patocchi Research Station, Agroscope Changins-Wädenswil, Wädenswil, Switzerland

Andreas Naef Research Station, Agroscope Changins-Wädenswil, Wädenswil, Switzerland

Anne Vanasse Département de phytologie, Faculté des sciences de l'agriculture et de l'alimentation, Département des sciences animales, Université Laval, Québec, Province de Québec, Canada

Aude Alaphilippe INRA, UERI Domaine de Gotheron, Saint Marcel-lès-Valence, France

Bart Heijne Applied Plant Research, Wageningen UR, Zetten, The Netherlands

Benoit Sauphanor INRA, UERI Domaine de Gotheron, Saint Marcel-lès-Valence, France

Boelie Elzen Wageningen UR Livestock Research, Lelystad, The Netherlands

Börje Johansson Farmer, Hulta Norrgård, Linköping, Sweden

Carlos Fonseca Centre for Transdisciplinary Development Studies (CETRAD), University of Trás-os-Montes and Alto Douro (UTAD), Vila Real, Portugal

Carlos Iorio Facultad de Ciencias Agrarias, Universidad Nacional de Mar del Plata Argentina, Buenos Aires, Argentina

Christian Thalmann School of Agricultural, Forest and Food Sciences, Bern University of Applied Sciences, Zollikofen, Switzerland

Claire Lavigne INRA, UERI Domaine de Gotheron, Saint Marcel-lès-Valence, France

Claudia R. Binder Department of Geography, LMU University, Munich, Germany

Diane Parent Département des sciences animales, Faculté des sciences de l'agriculture et de l'alimentation, Université Laval, Québec, Province de Québec, Canada

Didier Bébin INRA, Saint-Genès-Champanelle, France

Doris Pellerin Département des sciences animales, Faculté des sciences de l'agriculture et de l'alimentation, Université Laval, Québec, Province de Québec, Canada

Elena Bulfoni Department of Agricultural and Environmental Sciences, University of Udine, Udine, Italy

Elias Ghadban School of Oriental and African Studies, Department of Development studies, University of London, London, UK

Emiliana Silva Centre of Applied Economics Studies of the Atlantic (CEEAplA), University of Azores, Angra do Heroísmo, Açores, Portugal

Esther Bravin Research Station, Agroscope Changins-Wädenswil, Wädenswil, Switzerland

Filipa Manso Centre for Mountain Research, University of Trás-os-Montes and Alto Douro (UTAD), Vila Real, Portugal

Francesco Danuso Department of Agricultural and Environmental Sciences, University of Udine, Udine, Italy

Franco Rosa Department of Food Sciences, University of Udine, Udine, Italy

Frank Hayer Research Station, Agroscope Reckenholz-Tänikon, Zürich, Switzerland

Franz Bigler Research Station, Agroscope Reckenholz-Tänikon, Zürich, Switzerland

Gabriele Mack Research Station, Agroscope Reckenholz-Tänikon, Zürich, Switzerland

George Zervas Department of Nutritional Physiology and Feeding Faculty of Animal Science and Aquaculture, Agricultural University of Athens (AUA), Athens, Greece

Gérard Gaillard Research Station, Agroscope Reckenholz-Tänikon, Zürich, Switzerland

Giuseppe Feola Department of Geography and Environmental Science, University of Reading, Reading, UK

Guy Allard Département de phytologie, Faculté des sciences de l'agriculture et de l'alimentation, Département des sciences animales, Université Laval, Québec, Province de Québec, Canada

Heinrich Höhn Research Station, Agroscope Changins-Wädenswil, Wädenswil, Switzerland

Jan Grenz School of Agricultural, Forest and Food Sciences, Bern University of Applied Sciences, Zollikofen, Switzerland

Jean Luc Favreau National School of Agronomic Training, UMR Dynamiques Rurales, Castanet-Tolosan, France

Jesus Avilla Centre UdL-IRTA for R+D, AGROTECNIO, University of Lleida, Lleida, Spain

Joan Solé Centre UdL-IRTA for R+D, AGROTECNIO, University of Lleida, Lleida, Spain

Johanna Björklund, Ph.D. Man-Technology-Environment, School of Science and Technology, Örebro University, Örebro, Sweden

Jörg Samietz Research Station, Agroscope Changins-Wädenswil, Wädenswil, Switzerland

Jörn Strassemeyer JKI; Julius Kühn-Institute, Federal Research Centre for Cultivated Plants, Kleinmachnow, Germany

José Hernandez Research Station, Agroscope Reckenholz-Tänikon, Zürich, Switzerland

Joseph Le Blanc ADEAR-LR, Lattes, France

Konstantinos Tsiboukas Department of Agricultural Economics and Rural Development, Agricultural University of Athens (AUA), Athens, Greece

Luís Tibério Centre for Transdisciplinary Development Studies (CETRAD), University of Trás-os-Montes and Alto Douro (UTAD), Vila Real, Portugal

Mabelle Chedid Environment and Sustainable Development Unit, American University of Beirut, Beirut, Lebanon

Marijke Meul Department of Biosciences and Landscape Architecture, University College Ghent, Ghent, Belgium

Marko Bohanec Department of Knowledge Technologies, Jožef Stefan Institute, Ljubljana, Slovenia

Médulline Terrier CEMAGREF, UR DTM, Saint Martin d'Hères Cedex, France

Michel Lherm INRA, Saint-Genès-Champanelle, France

Mirna Mosciaro Estación Experimental Balcarce, Insituto Nacional de Teconología Agropecuaria (INTA) Argentina, Buenos Aires, Argentina

Mohamed Gafsi National School of Agronomic Training, UMR Dynamiques Rurales, Castanet-Tolosan, France

Patrick Veysset INRA, Saint-Genès-Champanelle, France

Patrik Mouron Research Station, Agroscope Reckenholz-Tänikon, Zürich, Switzerland

Pierre Gasselin INRA, UMR951 Innovation, Montpellier Cedex 1, France

Salma Talhouk Landscape Design and Ecosystem Management Department, American University of Beirut, Beirut, Lebanon

Shady K. Hamadeh Environment and Sustainable Development Unit, American University of Beirut, Beirut, Lebanon

Sierk F. Spoelstra Wageningen UR Livestock Research, Lelystad, The Netherlands

Steven Van Passel Centre for Environmental Sciences, Hasselt University, Diepenbeek, Belgium

Ursula Aubert Research Station, Agroscope Reckenholz-Tänikon, Zürich, Switzerland

Valérie Bélanger Département de phytologie, Faculté des sciences de l'agriculture et de l'alimentation, Département des sciences animales, Université Laval, Québec, Province de Québec, Canada

List of Figures

List of Tables

Chapter 1
The Needs for Building Sustainable Farming Systems: Issues and Scope

Emiliana Silva and Ana Alexandra Marta-Costa

Abstract Food sufficiency, environmental preservation, socio-economic viability and equity are the major components of sustainable farming. However, establishing the definitions and operational methodologies that enable their application in the decision-making process and in the sustainability assessment processes has proved to be a very difficult task. It is necessary to select, from among the various approaches, the methods of evaluation on which decision-making is based. In this chapter, there are briefly presented the needs for building farming systems enclosed in the main objectives of this book. In the final part of the chapter, the book organization and a short summary of the highlights for each of the chapters focusing on the methodologies used are presented.

Keywords Sustainable farming systems • Sustainability farming assessment • Methods • Book organization

1.1 Introduction

The environmental sustainability of farming systems is on the agenda of the European policy. The world citizens, including farmers, are increasingly more concerned regarding production respecting the environment, and they are changing some of their ways of production choosing more environmental friendly systems and technologies.

E. Silva (✉)
Centre of Applied Economics Studies of the Atlantic (CEEAplA),
University of Azores, 9700-042 Angra do Heroísmo, Açores, Portugal
e-mail: emiliana@uac.pt

A.A. Marta-Costa
Centre for Transdisciplinary Development Studies (CETRAD),
University of Trás-os-Montes and Alto Douro (UTAD), Quinta de Prados,
Apartado 1013, 5001-801 Vila Real, Portugal
e-mail: amarta@utad.pt

A.A. Marta-Costa and E. Silva (eds.), *Methods and Procedures for Building Sustainable Farming Systems: Application in the European Context*,
DOI 10.1007/978-94-007-5003-6_1,

The systems were thought as they behaved in a linear and predictable manner, but in reality, they are characterized by non-linearity, uncertainty, and prone to dramatic changes. Farmers and other stakeholders in rural areas need to develop the capacity to cope with, adapt and transform to these dynamics. Although there is much conversation of systemic approaches, their application to farming and rural settings is still starting. Indeed, the feedback loops in family farming systems and the dynamics of rural networks are still poorly understood.

In this context, the International Farming Systems Association (IFSA) promoted the 9th European IFSA Symposium, in Vienna, Austria, 2010, according to the theme "Building Sustainable Rural Areas", to discuss the themes: knowledge systems, learning and collective action; transition, resilience and adaptive management; energy production, CO_2 sink and climate change; sustainable food systems; and landscape and rural land use.

This book is a compilation of the revised and reworked papers presented in the Workshop 2.1 of Methods and Procedures for Sustainable Farming Systems in which the convenors were Alexandra Marta-Costa and Emiliana Silva. This workshop was included in the main theme Transition, Resilience and Adaptive Management. The theme intended to give answer to some important questions: what attitudes, structures and activities build and sustain the ability of farmers and farming communities to cope with ongoing change and what role of flexibility and diversity play in enabling transformation pathways and in enabling actors to take advantage of the opportunities offered by change.

Many authors have conducted their research into the requirements for sustainable farming, and most agree that food sufficiency, environmental preservation, socio-economic viability and equity are the major components of sustainable farming. However, establishing the definitions and operational methodologies that enable their application in the decision-making process and in the sustainability assessment process has proved to be a very difficult task. Still, it is urgent to arrive to a definition regarding technology solutions for a new farming productivity, within a framework of global sustainability, renewable use of natural resources and new technology, where product quality is maximized while the quality of landscape and rural life are simultaneously preserved.

In the current context, decision-making in the planning and management of farming activities is a complex – and potentially controversial – process, since it inevitably involves diverse environmental, social and economic objectives, which may or may not be mutually exclusive but which are clearly antagonistic, since the securing of one objective implies the underperformance of others.

For example, farming systems are faced with a double and generally contradictory challenge to be successful. On the one hand, invested capital has to be profitable and economic performance has to be maximized. On the other, given the socio-environmental situation, it is necessary to preserve and protect the environment and natural resources. From a normative standpoint, it is our view that farms should be planned and managed in a way that allows them to reach a compromise between the two principles – socio-economic sustainability and environmental sustainability.

However, choosing the best alternative is not sufficient: it is also necessary to select, from among the various approaches, the methods of evaluation on which decision-making is based.

This book is an opportunity to discuss the current research, especially by addressing following issues: (1) What methodologies are being developed for building sustainable farming systems? (2) What are their results? (3) Are there effective proposals to build sustainable farming systems efficiently? Consequently, this book and its 19 chapters focus on the methodologies used in the assessment of sustainability as a global concept (economic, social and environmental).

As noted above, the sustainability of agricultural farming is a preoccupation of the main European countries, and it seems that Europe can only have a way to have agricultural production: environmental friendly, without losing the economic and social vision of rural world.

1.2 Scope of the Book

Eighteen more chapters are compiled, grouped in seven parts. Part I intends to reveal the European context for sustainability assessment of farming systems and contains two chapters. Chapter 2 presents the European Union policies and their history in the ambit of relationship of agricultural and environmental themes. Chapter 3 is a synthetic presentation of the approaches used in this book for sustainable farming systems assessment.

Part II until Part VI aggregate the contributions of 60 authors from 12 countries that identify methods and procedures used or with potential of use for building sustainable farming systems or to assess the sustainability.

Part II has a focus on multiple aspects of procedural issues for sustainability assessment of farming systems. In Austria, Binder and Feola (Chap. 4) for evaluating sustainability assessment methods analysed their normative, systemic and procedural dimensions and applied it to indicator-based sustainability assessment methods in agriculture. Methods were categorized into three types: (i) top-down farm assessment methods, (ii) top-down regional assessment methods with some stakeholder participation and (iii) bottom-up integrated participatory or transdisciplinary methods with stakeholder participation. In France, Terrier, Gasselin and Le Blanc (Chap. 5) analysed three methods to appreciate the farm sustainability and identify their limits and their contributions to their methodology, at the level of complex activity systems in which farming production is combined with transformation, sales or outside activities. They proposed a farm-focused sustainability and an extended sustainability, which means a contribution to the sustainable development at a regional scale. In Belgium, Van Passel and Meul (Chap. 6) used a useful combination of methods with different purposes making sustainability assessment more profound and broadens the possible applications. In this, a combination of the Sustainable Value Approach (SVA) and Monitoring Tool for Integrated Farm Sustainability (MOTIFS) is presented. SVA is used to support policymakers, while MOTIFS is used to support farmers towards sustainability.

Three examples of application of sustainability assessment methodologies in real contexts are grouped in Part III. In the Netherlands, Elzen and Spoelstra (Chap. 7) developed two broad approaches, top-down and bottom-up. Currently, the links between the bottom-up and the top-down processes are relatively weak. As both may contribute to a system innovation, the major challenge is to make a fruitful combination between the two approaches, the "Learning and Experimentation Strategy" (LES). In Switzerland, Thalmann and Grenz (Chap. 8) used the Response-Inducing Sustainability Evaluation (RISE) as a method for rapid yet holistic sustainability assessment of agricultural production at farm level. Also in Switzerland, Mouron and nineteen more authors (Chap. 9) derived a multicriteria tool based on a hierarchical attribute tree, which uses qualitative ratings to rank results retrieved from life cycle assessment (LCA), the indicator model SYNOPS and full cost calculations. Results from a case study of 5 European countries being partners of the EU-FP6 project ENDURE demonstrate the feasibility to identify crop protection strategies with improved overall sustainability applying our new tool.

Particular aspects of the sustainability assessment in organic farming and multifunctional systems are exposed in Part IV. In France, Veysset, Lherm and Bébin (Chap. 10) used a model-based study for the conversion to organic farming, and it was simulated for three suckler cattle farms by coupling an economic assessment model with an environmental assessment. In Sweden, Björklund and Johansson (Chap. 11) used a holistic approach to measure efficiency, developed by participatory research. In their study, energy analysis and footprinting were combined to assess and illustrate the total resource use caused by a dairy farm. In France, Gafsi and Favreau (Chap. 12) proposed a method for assessing farm's sustainability suitable for organic farming taking into account the agro-ecological and socio-territorial specificities of this farming system.

Part V is dedicated to the decision support methods for sustainable farming systems. In Italy, Rocca, Danuso, Rosa and Bulfoni (Chap. 13) used a dynamic simulation to help in improving the farm management. This *X-farm* is composed by modules representing the farm activities and the main centres of farming costs: soil, management, crop production and processing, energy production and administration. In Portugal, Marta-Costa, Manso, Tibério and Fonseca (Chap. 14), given the multidimensional nature of sustainability as a result of interaction and complementarity between the economic, social and environmental dimensions, carried out the planning of a farming system having for base the multiobjective programming. Sintorini, Tsiboukas and Zervas (Chap. 15) proposed a whole-farm optimization model used to assess the socio-economic and environmental performance of the sheep farming activity in Greece. The analysis is undertaken in two sheep farms that represent the extensive and the semi-intensive farming systems.

Procedures and methods for sustainability assessment used in non-European farming systems are grouped in Part VI. In Canada, Parent, Bélanger, Vanasse, Allard and Pellerin (Chap. 16) proposed the assessment of farm sustainability based on its economic, environmental and social aspects. A holistic method, named DELTA, was developed for these three aspects. To identify the indicators, they used a multiple stakeholder perspective (researchers, farmers, advisors). Ghadban, Talhouk, Chedid and Hamadeh (Chap. 17) modified a French agriculture sustainability assessment model (IDEA-Indicateur de durabilité des exploitations agricoles) to fit the Lebanese agriculture context. The IDEA model was structured around three sustainability scales: agro-ecological, socio-territorial and economic scale translated into measurable indicators. In Argentina, Mosciaro and Iorio (Chap. 18) analysed the productive strategies of farms with intermediate levels of capitalization to infer the resource allocation tendency. The analysis incorporates market and production risk considerations through application of two Minimization of Total Absolute Deviations (MOTAD) models.

The final part (Part VII) is dedicated to the final considerations of the book. It was targeted to answer the main questions drawn up with the scientific knowledge treated in the book.

Part I
European Context for Sustainability Assessment of Farming Systems

Chapter 2
Agricultural and Environmental Policies in the European Union

Emiliana Silva and Ana Alexandra Marta-Costa

Abstract This chapter intends to display the environmental considerations in the evolution of agricultural policies of the European Union. In the initial part, it is stated that the agricultural policies are changing. The Common Agricultural Policy (CAP) of the European Union appears as the result of solving the problem for the food decrease production after the Second World War. Nowadays, the CAP primary principles and objectives are changing according to the globalisation and the European environmental concerns. CAP has adjusted to a new reality – the sustainability of the agricultural production. As consequence, the CAP has evolved significantly since 1950 and developed important measures. The common market organisation and other policies including environmental concerns pointed in the MacSharry Reform, the check-health and in the agri-environmental measures appeared as an alternative of the old CAP. The future CAP will have also environmental objectives. CAP will fight against climate changes, supporting employment and growth; promoting the environmental protection and rural development plans and food security; and avoiding global hunger and poverty.

Keywords Common agricultural policy • Agri-environmental measures • Environmental policy • European Union

E. Silva (✉)
Centre of Applied Economics Studies of the Atlantic (CEEAplA),
University of Azores, 9700-042 Angra do Heroísmo, Açores, Portugal
e-mail: emiliana@uac.pt

A.A. Marta-Costa
Centre for Transdisciplinary Development Studies (CETRAD),
University of Trás-os-Montes and Alto Douro (UTAD), Quinta de Prados,
Apartado 1013, 5001-801 Vila Real, Portugal
e-mail: amarta@utad.pt

A.A. Marta-Costa and E. Silva (eds.), *Methods and Procedures for Building Sustainable Farming Systems: Application in the European Context*,
DOI 10.1007/978-94-007-5003-6_2,

2.1 Introduction

The agriculture is one of the main sectors of the European Union (EU) economy to obtain the attention of the policymakers through its Common Agricultural Policy (CAP). This is due to the people employed in the agricultural sector and that most of the EU surface is used in agriculture. However, this sector has a significant effect in environment. The European land is mainly covered by farms and forests and its management promotes serious impacts on the atmosphere, water, soil and other natural resources. Agriculture is also essential for the health and economy (inclusive food production) (EUROSTAT 2010).

The CAP has evolved significantly since is appearance. First of all, it was mainly focused in the agricultural production and in its subsiding sector, after successfully controlling the policy production (overproduction of the supply). The actual CAP aims that farming and environmental preservation drive simultaneously. Nowadays, it is associated to food safety and market organisation; it helps the development of the economic and social rural communities and has a major environmental importance in the new challenges such as climate change, water management, bioenergy and biodiversity (EU 2011e).

The CAP is now in the agenda of EU and there was a large debate in the last year, regarding the future of the agricultural EU policy. This chapter aims to present the evolution of CAP, since is appearance to the recent years. It starts with the principles and initial objectives of CAP and is followed by the most important measures involving the environment – the MacSharry reform, the check-health and the agri-environmental measures. In the final part, it points out the future of CAP.

2.2 The CAP Evolution

The CAP appears (in the 1950s and 1960s) as the result for solving the decrease of food production problem after the Second World War. Most part of the European countries was completely destroyed, as well as their agricultural productions. The market was uncontrolled and the supply was lesser than the demand of food production. So, the agricultural policy occupies a major role in the economic and policy in the new EU institution.

Initially, the main objectives were concerned to produce agricultural products and eliminate the tariffs between the partnership countries. To achieve this, it was agreed to follow five objectives of the article 39 of the Treaty of Rome, in 1957 (Pezaros 1998):

1. To increase agricultural productivity, by promoting technical progress and ensuring the optimum use of the production factors, in particular labour
2. To ensure a fair standard of living for the agricultural community
3. To stabilise markets
4. To secure availability of supplies
5. To provide consumers with food at reasonable prices

These objectives promoted the three next principles:

1. Single market (a common customs tariff, rules on internal competition and institutional prices and aids)
2. Community preferences (products from EU have priority on the internal market access and a protection against similar products from others countries)
3. Financial solidarity (the organisation of markets was totally financed by the community budget)

As outcome of the success of the agricultural policies, some problems, distorting trade and environmental concerns, appeared. These situations had as consequence the successive changes of the CAP along the last decades.

The CAP is organised into two pillars: pillar (1) the market policy and pillar (2) the sustainable development of rural policy. The pillar 1 includes the common market organisation (for instance, the direct payment) and the pillar 2 the rural development regulations (including the national or regional development plans).

Some measures, supported by the first pillar, are the tariffs in the imported goods, exports licences and refunds, the quotas, the intervention mechanisms and others measures such as the EU milk scheme (OECD 2011). The second pillar of the CAP includes measures such as farm modernisation, the setting up of young farmers, early retirements, vocational training, the leader approach, agri-environmental and animal welfare and so on. Nowadays, the organisation of the CAP is still grounded in two pillars but giving an increasing importance for the second pillar policies in the agricultural budget.

One of the starter tools of CAP was achieved by a Common Organisation of Markets (COM) based on market prices support system. It started with cereals (in 1962) but nowadays almost all agricultural products have one. The COM is a set of rules and institutional measures, including the cycle of production, processing and marketing of agricultural products and the differentiation of prices (institutional, target or guide price; support or intervention price; and an imported price) withdrawals, production aids, storage, quotas (limits of productions), export refunds (subsidy to promote the exportation), import levies (to decrease the importations) and others.

The integration of the environment into the agricultural policy began in the 1980s. The Gundelach report (from 1980) and the Thorn report (from 1981) can be considered as the documents that started the revision process culminating in the MacSharry reform (from 1992) (Blasi et al. 2007). The main measures in the reform were based in the changing of system prices and involving the concept of separation of production quantity and market orientation. The financial pressures led to wholesale EU agricultural reforms in the 1980s and 1990s, the 'MacSharry plan' measures, through the introduction of production quotas, aimed for the realignment with the world markets and the decoupling of commodity support from tonnage-produced to acreage planted, thus effectively capping the agricultural budget (Gibbard 1997). The MacSharry plan is a social plan involving the redistribution of support targeted towards smaller farms, thereby reorienting the CAP socially and economically to enable a significant number of families to remain on the land. This plan intends to end with a higher productive CAP. For instance, some policies in the MacSharry reform were early retirement, agri-environmental measures and afforestation scheme.

After the 1992 MacSharry reform, direct payments to EU farmers were introduced and become an integral part of the CAP, like a farmers' incomes compensation as consequent of losses for the decreasing of intervention prices (Santos et al. 2010). The main measures of the reform of 1992 were:

- Aid for the modernization of holdings
- Installation of young farmers
- Aids for processing and marketing
- Diversification of production

These reforms had as goal to reduce support prices and to complement the farmers' incomes with a direct aid payment.

In the 1990 decade, the changes of CAP aimed to support the production towards a market-oriented and agricultural sustainability (more environmental friendly practices). The Treaty of Amsterdam (in 1997) reinforces sustainable development as an objective of the EU and the sustainable agriculture needs to reflect productive, environmental and social functions. That is:

- Agriculture is an economic activity and should guarantee food security but also to provide higher-quality food and non-food products responding to the new consumer requirements.
- Agriculture plays an increasingly important role in the environment sustainability.
- Agriculture is an important factor contributing to a balanced economic and social rural communities defined by a rural development policy.

The changes introduced in 1992 were consolidated with the CAP reform of Agenda 2000. This new reform aimed to improve the balance between agriculture and environment. The main issues were (1) to apply compulsory restrictions, (2) to apply cross-compliance and (3) to use agri-environmental programmes (Santos et al. 2010).

Also, the Agenda 2000 shifts the price support to direct income. The 'direct payment' (single farm payment), an important step in 2003 and 2004, was supposed to be a system decoupling support from production and, also, connected to the consumers concern. The whole issue played a decisive role in the course of the general agreement tariffs and trade multilateral negotiations, as well as in the context of the internal discussions held on the orientations of 1992 reform. In these last discussions, the 'price-support system' was accused of causing surpluses and trade distortion since it was linked to the volume of production and trade.

All these policies intend to encourage entrepreneurial behaviour. As a consequence, the farmers can react better to the markets to introduce improved techniques; to promote diversified activities such as rural crafts, food processing facilities on farms, tourism or forestation; to promote environmentally friendly farming practices; and to use other rural development tools.

One of the most important measures reforms of agricultural policy was a health check (in 2007). This policy aimed to prepare the CAP to financial framework for

2013 in the context of pillar 2 (Santos et al. 2010). It was an opportunity to ensure that the policy is fit for new challenges and opportunities, such as climate change.

A health check policy had as objective assessing the implementation of 2003 CAP reform and introduce new adjustments: to modernise, to simplify, streamline the CAP and to remove restrictions on farmers, helping them to better answer to the market and to new challenges (EU 2011a). For instance, one measure was to increase milk quotas gradually leading to their abolition in 2015 and to use market intervention as a safety net when food prices are very low. Also, an important step in the health check was the reduction of direct payments to farmers and the money transferred to a fund for the development of rural regions, which means a shift in funding from direct payment to a rural development support. Other measures were the agreement to abolish arable set-aside and the conversion market intervention into a safety net. It is expected an adequate response to the new challenges and opportunities faced by European agriculture, including climate change, better water management, protection of biodiversity and green energy production.

2.3 The Engagement of Environmental and Agricultural Policies

The CAP has been increasingly adapted for integrating environmental concerns and to serve best the sustainability purposes. This adjustment is based on a distinction between ensuring a sustainable way of farming by avoiding environmentally harmful agricultural activity and providing incentives for environmentally beneficial public goods and services. For ensuring sustainable agricultural activities, farmers have to respect rules and standards for preserving the environment and the landscape.

To provide a better environment, farmers have to voluntarily or compulsorily respect legislation. In the first case, there are appropriate incentives. Farmers are remunerated for voluntarily engaging in environment activities, this is the provider gets principle. The other principle of CAP is the polluter pays principle (EU 2011d and CEC 2000).

The polluter pays principle states that the polluter should bear the costs of avoiding or remedying environmental damage. Generally, farmers have to ensure compliance with mandatory national and European environmental standards and respect the basic mandatory standards forming part of the cross-compliance regime at their own costs. Non-compliance with mandatory requirements is subject to sanctions. It implies a ‘good farming practice’ as a relationship between agriculture and environment if they are rightly compensated.

The provider gets principle is described as remunerating voluntary environmental commitments going beyond legal requirements. For the CAP, this principle is taken up via agri-environment payments which encourage farmers to sign up for environmental commitments beyond the reference level of mandatory requirements. Agri-environment payments shall cover the costs incurred and income forgone as resulting from voluntary environmental commitments.

The two mechanisms to integrate the environment in agriculture are:

1. The cross-compliance to most CAP payments and sanctioning non-compliance by payment reductions (pillar 1)
2. The agri-environment measures (pillar 2)

The cross-compliance, incorporated in the horizontal regulation, links with direct payments to compliance by farmers with basic standards concerning the environment, food safety, animal and plant health and animal welfare, as also the requirement of maintaining land in good agricultural and environmental condition (EU 2011b; Lacroix 2003 and Commission of European Communities – CEC 2000). The cross-compliance policy includes two elements:

1. Statutory management requirements: these requirements refer to the field of the environment, food safety, animal and plant health, and animal welfare
2. Good agricultural and environmental condition: refers to a range of standards related to soil protection, maintenance of soil organic matter and structure, avoiding the deterioration of habitats and water management

The cross-compliance represents the 'reference level' for agri-environment measures. The limit is the polluter pays principle. The cross-compliance promotes the support granted under the CAP, contributes for promoting sustainable agriculture and, also, is near to the concerns of the European citizens.

The environmental integration in agriculture is recognised by four measures (EU 2011c):

1. Market stability or income support could influence the positive effects on the environment or contribute to maintaining environmentally beneficial structures or types of farming (less favoured areas – areas of the EU where natural physical conditions cause lower agricultural productivity – payments)
2. Income support, for contributing to the enforcement of mandatory environmental requirements and the polluter pays principle (decoupled payments in combination with cross-compliance)
3. Promoting the provision of environmental voluntary services (agri-environment measures)
4. Facilitating compliance with compulsory environmental requirements (meeting standards measure) or compensate the relative economic disadvantage resulting from a region-specific pattern of environmental requirements (water framework directive)

The agri-environment measures provide payments to farmers who voluntarily subscribe to environmental commitments related to the preservation of the environment and maintaining the countryside. These payments provide compensation for additional costs and income foregone resulting from applying environmentally friendly farming practices. Farmers have a period of at least 5 years to adopt environmentally friendly farming techniques.

Some examples of agri-environmental schemes are environmentally favourable extensification of farming; management of low-intensity pasture systems; integrated farm management and organic agriculture; landscape preservation and historical features such as hedgerows, ditches and woods; and conservation of high-value habitats and their associated biodiversity. Agri-environment measures have an important role for meeting the society's demand for environmental outcomes provided by agriculture.

2.4 The Future CAP

On April 2010, in EU, a public debate on the CAP's future, objectives, principles and contributions to the 'Europe 2020' strategy was held. This debate was centred in around four main questions (EU 2011g):

1. Why do we need a European CAP?
2. What are society's objectives for agriculture in all its diversity?
3. Why should we reform the CAP and how can we make it meet society's expectations?
4. What tools do we need for tomorrow's CAP?

The three scenarios encountered in the impact assessment were (European Commission 2011):

- Scenario 1 – an adjustment scenario that continues with the current policy framework while addressing its most important shortcomings, such as the distribution of direct payments
- Scenario 2 – an integration scenario that entails major policy changes in the form of enhanced targeting and greening of direct payments and reinforced strategic targeting for rural development policy in better coordination with other EU policies, as well as extending the legal base for a broader scope of producer cooperation
- Scenario 3 – a refocus scenario that reorients the policy exclusively towards the environment with a progressive phasing out of direct payments, assuming that productive capacity can be maintained without support and that the socio-economic needs of rural areas can be served by other policies

The major problems identified and pointed in the actual CAP were:

- The EU agricultural tariffs and subsidies distort the economy
- The CAP harms EU trade interests
- The CAP is socially unfair
- The CAP has a weak environmental record
- The CAP undermines global food security and the fight against poverty
- The CAP is a burden on European integration

The current reform proposals supported by the inter-institutional debates and by stakeholder consultation were identified on the Communication on the CAP towards 2020 (CEC 2010) and have as objectives (1) viable food production, (2) sustainable management of natural resources and climate action and (3) balanced territorial development.

These purposes for CAP were also formulated by the CEC, in the International Centre for Trade and Sustainable Development (ICTSD 2011). In this document, it is stated that future reform of the European agricultural sector (CAP) aims to strengthen the competitiveness, the sustainability of agriculture, and maintain its presence in all 27 countries belonging to the EU. This institution proposes a new partnership between Europe and their farmers in order to guarantee European citizens healthy and quality food production, to preserve the environment and to help develop rural areas. In the document of ICTSD (2011), Dacian Cioloş refers the next decades as crucial for positioning the strong basis for an agricultural sector that can cope with climate change, international competition and the expectations of the citizens.

The economic efficiency and competitiveness requested for the future CAP include also rural public goods. The future role of the CAP should be to give farmers appropriate incentives to deliver European public goods demanded by the society, particularly related to the environmental issues. This includes the fight against climate change, the protection of biodiversity and the water management (avoiding pollution, scarcity and floods) (EU 2011h).

As synthetised, the Group of Leading Agricultural Economists (in declaration 2009, EU 2011h) identified four classes of potential objectives for the future CAP:

1. Enhancing economic efficiency and competitiveness
2. Ensuring food security
3. Changing income distribution
4. Promoting public goods

However, for this group, only the public goods can provide a sustainable basis for the future CAP. For Zahrnt (2011a, b), the main issue for upcoming CAP is centred in the food security concerns, related to the size of EU production, to its production potential and the reliability of imports. The global food security or world hunger is a serious concern.

Naylor (2011) suggests that the broader issues of international and human welfare will be incorporated in the future of agricultural policy as a way to reduce the global hunger and poverty. This means that the actual financial crises and the economic, social and environmental context must restrain the agricultural policy in a different standard. International policy needs to embrace a wider perspective on agricultural development to avoid food insecurity.

The ten key points of the future CAP reform (2013–2020), based on documents of RAPID (2011) and EU (2011f), are the following:

1. Better targeted income support in order to stimulate rural growth and employment. To better develop the agricultural potential of the EU, it was proposed to support farmers' income in a fairer, better targeted and simpler way. Basic income support will cover only active farmers. It will be digressive from € 150 000,00 per holding and capped beyond € 300 000,00, taking into account the number of jobs created. It will also be distributed more equitably between farmers, regions and European countries.
2. Tools to address crisis management which are more responsive and better suited to meet the economic challenges. Price volatility is a threat to the long-term

competitiveness of the agricultural sector. This can be achieved by proposing safety nets which are more effective and responsive for the sectors more exposed and, also, to promote the creation of insurance and mutual funds.

3. A 'green' payment for preserving long-term productivity and ecosystems. To strengthen the environmental sustainability of agriculture and enhance the efforts of farmers, it is proposed to spend about 30% of direct payments for the improved use of natural resources. Some measures such as crop diversification, maintenance of permanent pasture and preservation of environmental reservoirs and landscapes are practical, simple to implement and will have an environmental effect.
4. More investment in research and innovation. To produce efficiently, it is proposed to double the budget for agricultural research and innovation. These funds will support research projects with importance to farmers, to promote closer cooperation between researchers and farmers and the good communication of results from the research institute and the farms, and provide better information and help to farmers.
5. A more competitive and balanced food chain. The agriculture is the first step in the food supply chain, but the sector is highly fragmented and unstructured, and its added value is not recognised. To strengthen the position of farmers, the CAP will support producer organisations, develop inter-professional organisations and develop direct sales between producers and consumers.
6. Agri-environmental initiatives. The specificities of each member state and its regions (including the ultraperipheral regions) should be taken into account, and environmental practices must be encouraged at national, regional and local level. The tools are the rural development policy priorities for restoring, preserving and enhancing ecosystems and for resource efficiency and less influence of climate change.
7. Facilitating the establishment of young farmers. To help the younger generation to get involved in the agricultural sector, it will be creating a new installation aid available to farmers under 40 years old during the first 5 years of their project.
8. Stimulating rural employment and entrepreneurship. To promote employment and entrepreneurship, for example, a 'starter kit' will be created to support microenterprise projects with funding up to € 70 000,00 over 5 years. The leader local action groups will be strengthened.
9. Better addressing fragile areas. To avoid desertification and preserve the richness of the European lands, it is providing an opportunity for the member states to further help farmers in areas with natural handicaps, with additional support, like Azores, in Portugal. It will be adding to other aid already available under the rural development policy.
10. A simpler and more efficient CAP. It will be simplifying some administrative mechanisms of the CAP, including the rules of conditionality and control systems, and the aid to small farmers will also be simplified. For the latter, a flat rate of € 500,00 to € 1 000,00 per farm per year will be created. The sale of land by small farmers who cease agricultural activity to other farms willing to restructure their farms will be promoted.

To solve the budget problem of the CAP, it should be significantly reduced. The first pillar of the CAP should be progressively abolished; many policies under the second pillar should be removed and, finally, must avoid subsidies that distort competition or harm the environment. The objectives of the CAP should be, mainly, environmental objectives, and its budget must be spent on European public goods (EU 2011i).

For balance agriculture and environment in CAP reform, there are three identified target areas (EU EU 2011g):

1. Biodiversity, preservation and development of 'natural' farming and forestry systems and traditional agricultural landscapes
2. Water management and use
3. Dealing with climate change

2.5 Final Remarks

Since 1992, the agricultural policy is trying to mitigate the CAP's negative effect, namely, in the environment. Successive reforms of CAP in EU had been developed to reduce the support and to change the way that it has been carried out for and by farmers.

The MacSharry reform was important to promote a modernised CAP, recognising the role of agriculture expected by society and simultaneously involving the environmental protection and rural development. It promoted a better balance in the market (first pillar) and structural (second pillar) policies as Blasi et al. (2007) states.

Actually, the CAP needs a restructuration and reorganisation. The world and the EU are changing, the socio-economic and environmental contexts are in a dynamic process and the objectives and final products required by society for agrarian sector are being mutated. The integration of environmental considerations in agricultural policies is essential and urgent and sustainability is now an always present prerequisite.

The recent debate on CAP post-2013 represents an opportunity of the EU to build upon the considerable success of past policies reforms and to align future policy tools with its future objectives as stated in the document of OECD (2011). It is forecasted that the CAP will be more oriented to the environmental issues and to the second pillar of CAP (rural development issues).

References

Blasi, G., Boni, A., Gigante, R., & Roma, R. (2007). Evolution of agricultural income in Puglia between Mac Sharry Reform and the Mid-term review. *New Medit - A Mediterranean Journal of Economics, Agriculture and Environment, 3*, 42–53.

CEC. (2000). *Indicators for the integration of environmental concerns into the Common Agricultural Policy*. Communication from the Commission to the Council and the European Parliament, COM(2010)20 final, 26.01.2000. Brussels: CEC.

CEC. (2010). The CAP towards 2020: Meeting the food, natural resources and territorial challenges of the future. Communication from the Commission to the European Parliament, the Council, the European Economic and Social Committee and the Committee of the Regions, COM(2010)672 final, 18.11.2010. Brussels: CEC.

EU. (2011a). *"Health Check" of the Common Agricultural Policy*. Retrieved December 13, 2011, from http://ec.europa.eu/agriculture/healthcheck/index_en.htm

EU. (2011b). *Cross-compliance. European Union*. Retrieved December 13, 2011, from http://ec.europa.eu/agriculture/envir/cross-compliance/index_en.htm

EU. (2011c). *Agri-environment measures. European Union*. Retrieved December 13, 2011, from http://ec.europa.eu/agriculture/envir/measures/index_en.htm

EU. (2011d). *Integrating environmental concerns into the CAP*. European Union. Retrieved December 13, 2011, from http://ec.europa.eu/agriculture/envir/cap/index_en.htm

EU. (2011e). *Agriculture and the environment: Introduction. European Union*. Retrieved December 13, 2011, from http://ec.europa.eu/agriculture/envir/index_en.htm

EU. (2011f). *Legal proposals for the CAP after 2013*. European Union. Retrieved December 13, 2011, from http://ec.europa.eu/agriculture/cap-post-2013/legal-proposals/index_en.htm

EU. (2011 g). *The Common Agricultural Policy after 2013*. European Union. Retrieved December 13, 2011, from http://ec.europa.eu/agriculture/cap-post-2013/index_en.htm

EU. (2011 h). *A Common Agricultural Policy for European Public Goods. 2009. Declaration by a Group of Leading Agricultural Economists*. European Union. Retrieved January 7, 2012, from http://www.reformthecap.eu/declaration-2009

EU. (2011i). *CAP Reform in a Nutshell*. European Union, Retrieved January 7, 2012, from http://www.reformthecap.eu/cap-reform-in-a-nutshell

European Commission. (2011). *Regulation of the European Parliament and of the council amending Council Regulation (EC) no 73/2009 as regards the application of direct payments to farmers in respect of the year 2013*. Proposal draft, Brussels.

EUROSTAT. (2010). *The Eurostat yearbook, Europe in figures — Eurostat yearbook 2010*. EUROSTAT. Luxembourg: Publications Office of the European Union.

Gibbard, R. (1997). *The relationship between European Community agricultural structural policies and their implementation and agricultural succession and inheritance within Member States*. Reading: University of Reading, UK.

ICTSD. (2011). EU Farm Policy: Commission reform proposals spark mixed response. *Bridges weekly trade News Digest, 15*(34), 8–9.

Lacroix, E. (2003). *Agriculture and the environment*. Brussels: European Commission.

Naylor, R. (2011). Expanding the boundaries of agricultural development. *Food Security, 3*, 233–251.

OECD. (2011). *Evaluation of agricultural policy reform in the European Union*. OECD Publishing. Retrieved February 27, 2012, from http://dx.doi.org/10.1787/9789264112124-en

Pezaros, P. (1998). *An introduction to the Common Agricultural Policy: Principles, objectives and orientations*. Greece (1988-June 1997). Maastricht: CIHEAM – Options Mediterraneennes, European Institute of Public Administration. Retrieved October, 2011, from http://ressources.ciheam.org/om/pdf/c29/CI020491.pdf

RAPID. (2011). *A Comissão Europeia propôs uma reforma da PAC que vai fortalecer a agricultura Portuguesa. Edição da Representação da Comissão Europeia em Portugal*. Retrieved December, 2011, from http://europa.eu/rapid/pressReleasesAction.do?reference=IP/11/1181&format=HTML&aged=1&language=EN&guiLanguage=en

Santos, M., Fragoso, R., Henriques, P., & Silva, M. (2010). The agricultural competitiveness of the Alqueva Irrigation Project under the 2003 CAP reform and the health check. *New Medit, 4*, 11–18.

Zahrnt, V. (2011a). *Food security and the EU'S Common Agricultural Policy: Facts against fears* (ECIPE Working Paper, No. 01/2011). Retrieved January 7, 2012, from www.reformthecap.eu

Zahrnt, V. (2011b). A *guide to CAP reform politics: Issues, positions and dynamics* (ECIPE Working Paper No. 03/2011). Retrieved January 7, 2012, from www.reformthecap.eu

Chapter 3
Approaches for Sustainable Farming Systems Assessment

Ana Alexandra Marta-Costa and Emiliana Silva

Abstract Evaluating sustainable development is, at present, an essential prerequisite for promoting sustainable agriculture. The approaches for building sustainable farming systems presented in the next chapters of this book use various methods and procedures to assess the sustainability of farming systems. In that sense, this chapter presents some useful elements regarding those initiatives: Arbre de l'Exploitation Agricole Durable (ARBRE), Framework for the Evaluation of Sustainable Land Management (FESLM), Indicateur de Durabilité des Exploitacions Agricoles (IDEA), Indicator of Sustainable Agricultural Practice (ISAP), Multiscale Methodological Framework (MMF), diagnostic de durabilité du Réseau de l'Agriculture Durable (RAD), Response-Inducing Sustainability Evaluation (RISE), Sustainability Assessment of the Farming and the Environment (SAFE), and the Sustainability Solution Space for Decision Making (SSP) method.

Keywords Sustainable farming systems • Sustainability farming assessment • Sustainable farming methods

A.A. Marta-Costa (✉)
Centre for Transdisciplinary Development Studies (CETRAD),
University of Trás-os-Montes and Alto Douro (UTAD), Quinta de Prados,
Apartado 1013, 5001-801 Vila Real, Portugal
e-mail: amarta@utad.pt

E. Silva
Centre of Applied Economics Studies of the Atlantic (CEEAplA),
University of Azores, Rua Capitão João d'Ávila,
9700-042 Angra do Heroísmo, Açores, Portugal
e-mail: emiliana@uac.pt

A.A. Marta-Costa and E. Silva (eds.), *Methods and Procedures for Building Sustainable Farming Systems: Application in the European Context*,
DOI 10.1007/978-94-007-5003-6_3, © Springer Science+Business Media Dordrecht 2013

3.1 Introduction

Discussion about development concepts has been converted into an actual and interesting theme from theoretical and technical points of view and also from environmental, economic, social, and political gambits. These are necessary significant contributions to design conceptual structures and practical tools that can enable transforming theoretical idealizations into concrete actions.

Evaluating sustainable development is, at present, an essential prerequisite for promoting sustainable agriculture. For this achievement, it is necessary to understand operational models to assess sustainability in a tangible way that explicitly reflects the environmental, social, and economic advantages and disadvantages of different strategies and production systems (Masera et al. 2000).

The approaches for building sustainable farming systems employed in the next chapters follow various methods and procedures that are necessary to understand. Most of these approaches are indicator-based methods and are integrative assessment approaches for agricultural systems structured in rigorous and complex frameworks. All include the economic, environmental, and/or social dimensions of sustainability. They are, in alphabetical order, the Arbre de l'Exploitation Agricole Durable (ARBRE, sustainable farm tree), the Diagnostic de Durabilité (sustainability diagnosis) of the Réseau de l'Agriculture Durable (RAD), the Framework for the Evaluation of Sustainable Land Management (FESLM), the Indicateur de Durabilité des Exploitacions Agricoles (IDEA, sustainability indicator of farms), the Indicator of Sustainable Agricultural Practice (ISAP), the Multiscale Methodological Framework (MMF), the Response-Inducing Sustainability Evaluation (RISE), the Sustainability Assessment of the Farming and the Environment (SAFE), and the Sustainability Solution Space (SSP) for Decision Making method.

This chapter presents a synthesis of these initiatives that used for sustainability assessment of farming systems. When possible, there is shown, for each procedure, the authors and a brief description of the main methodology.

3.2 Global View of Methods to Assess Sustainability of Farming Systems

The development of methodological alternatives aimed to assess sustainability has conceptual problems and gaps that prevent, for the time being, conclusive statements. To assess farming systems sustainability, Binder and Wiek (2007) and Smith and McDonald (1998) identified, in agricultural multi-functionality, the scales to adopt, the selection of appropriate indicators, the linkages and integration of indicators, and the application of the results as being the main problems for these evaluation methodologies. To assess sustainability requires interdisciplinary and integrated efforts addressing the analysis of environmental processes and socioeconomic phenomena and multi-criteria models based on qualitative and quantitative indicators, it being necessary to integrate temporal perspectives wider than those usually used in conventional assessments (Masera et al. 2000).

Despite the problems commonly found with the definition and application of methodologies for assessing sustainability, it appears that the concern and efforts for their development led to a broader perception and more detailed reality (Marzall 1999).

Sustainability assessments became an area of intense research on an international scale. Many initiatives have been developed to assess sustainability in farming systems. Hansen (1996) identifies two groups of methodologies according to his interpretation of sustainability. The first is based in a goal concept that interprets agrarian sustainability as an ideological approach. This concept was developed in response to environmental problems from agriculture with the objective of motivating alternative agrarian practices. The second methodological group has its basis in a system-oriented concept. Sustainability is viewed as a property of agriculture that satisfies a diversified group of objectives which should continue over time. This concept is based on the impacts on agriculture viability resulting from global changes. In this group are methods that use multiple qualitative and quantitative indicators.

Other arrangements for sustainability assessment methods were proposed by Masera et al. (2000), who grouped these initiatives into four categories, according to their structure and measurement methods: (1) sustainability indicators; (2) sustainability indexes; (3) reference systems; and (4) frameworks for sustainability assessment. The sustainability indicators are selected parameters that can be isolated or interconnected and reflect conditions of the analyzed systems. The sustainability indexes aggregate, or synthesize, in one numerical value the relevant information for system sustainability from various indicators. Other theoretical efforts characterize ecological sustainability in an ecosystem perspective. The natural ecosystems are defined as reference systems to which management systems should attend. In the last group of sustainability assessment methodologies, Masera et al. (2000) refer to the sustainability assessment framework. The conceptual and practical efforts of this category are qualitatively distinct from the other groups: they have a more complex and rigorous structure. They integrate elements from different evaluation strategies, because indicators and indexes elaborate to iterative and participative analysis of farming systems.

In this last group, Van Cauwenbergh et al. (2007) distinguish two kinds of structures: (1) structures based in a systemic approach, with indicators that describe key attributes (functions or general processes) of all systems; and (2) structures based on disciplinary content, promoting specific indicators that characterize individual parts (related to specific functions or processes) of the analyzed systems.

Also, the initiatives to assess farming sustainability can be categorized into three typologies, according to Binder and Feola (see Chap. 4, this volume): (1) top-down farm assessments focus on field or farm assessment; (2) top-down regional assessment addresses both on-farm and regional effects; and (3) bottom-up, integrated participatory or transdisciplinary approaches focus on a regional scale.

The first type has a clear procedure for measuring indicators and assessing the sustainability of the system. However, the low degree of participation can negatively affect implementation of the results. The top-down regional assessment assesses both on-farm and regional effects. This second group includes some participation to increase acceptance of the results, but they fail in the analysis of potential trade-offs. The third group of methodologies integrates stakeholders throughout the whole process, assuring acceptance of the results and increasing the probability of implementation of developed measures, and also allow for performing trade-off analysis, because these methods include interaction between indicators in their system representation.

3.2.1 Arbre de l'Exploitation Agricole Durable (ARBRE)

ARBRE methodology was developed by TRAME, a federation of four non-profit associations in France. The sustainable farm tree is an aid-decision tool based on a qualitative approach and, if possible, a collective use. It aim is to help farmers build a business project on their farms, according to sustainable development stakes (Pervanchon 2007).

ARBRE is based on a set of about 60 qualitative questions corresponding to the dimensions of sustainable development: economy, transmission of capital and knowledge, social aspects, and environment. All answers are symbolized by a tree: the economic dimension is its stem. The tree symbolizes the fact that economy is a pillar for farmers. The liveability and social aspects in the roots demonstrate that not only family, but also social contacts, exchanges, and discussions with local or national partners, bring life to the farm. Environment is in the branches: it symbolizes what gives the farm its shape and what is seen from outside. Transmission is the fruits and leaves, what is collected, and what will make other trees, and the territory is the soil, from which the tree pumps its water and where the fruits and leaves, or organic matter, return (Pervanchon 2007).

3.2.2 Diagnostic de durabilité of Réseau de l'Agriculture Durable (RAD)

The Diagnostic de Durabilité of the RAD, developed in the far west of France by dairy farmers, is an evaluation method for setting targets and monitoring farm sustainability. It summarizes three types of sustainability assessment (IDEA, Solagro, and Fadear) [RAD and CIVAM (Centre d'Initiatives pour Valoriser l'Agriculture et le Milieu rural) 2010].

This tool is a diagnosis of sustainability used as an educational tool for self-evaluation and to aid reflection. This method is based on 22 indicators across three centers of interest: economic, social, and environmental sustainability. The set of criteria can be found in the accounts: balance, income account, manure plan, and loan amortization plan. The RAD methodology ends with three radars that offer the possibility to understand farm sustainability: the higher the surface, the more sustainable is the farm (RAD and CIVAM 2010).

3.2.3 Framework for the Evaluation of Sustainable Land Management (FESLM)

The FESLM is the Food and Agriculture Organization of the United Nations (FAO) framework to evaluate sustainable land management.

The FESLM pathway has two main stages. The first stage, with two levels, defines the purpose of the evaluation: what is to be evaluated. The second stage, with three levels, defines the process of analysis: how the evaluation is done. Level one identifies the land use system to be evaluated in terms of its purpose, its location, and the time period for sustainability. Level two defines the management practices to be employed to attain the objective. The qualities, attributes, processes, controlling interests, or constraints that affect sustainability in the context of the evaluation are identified on level three (second stage). Level four identifies how the selected evaluation factors impact sustainability through analysis of available information, modeling, expert systems, and experimentation. Level five identifies measurable or observable attributes that reveal the future status or condition of the evaluation factors and provide a measure of sustainability. In an "assessment endpoint" conclusions are drawn on the probable sustainability of the land use system as a whole. Then, the levels together require to be validated by reexamination of all the steps in the analysis to ensure that there has been consistency throughout the application of the framework principles and procedures (FAO 1993).

3.2.4 Indicateur de Durabilité des Exploitacions Agricoles (IDEA)

The IDEA method is a French self-assessment grid for farmers that provides operational content for the concept of agricultural sustainability. IDEA focuses in the preservation of natural resources and social values that are implicit in sustainable agriculture (Vilain 2008).

The matrix of the IDEA method is constructed with 41 indicators providing information on 16 objectives, grouped together to form three sustainability scales: (1) agro-ecological, (2) socio-territorial, and (3) economic sustainability. The first scale concerns the agronomic principles of integrated agriculture, which must enable good economic efficiency with an ecological cost as low as possible. The socio-territorial sustainability scale refers more to ethics and human development: these are essential features of sustainable agricultural systems. The last scale, economic sustainability, specifies the essential notions relating to the entrepreneurial function of the farm. The initial hypothesis of the IDEA method postulates that it is possible to quantify the various components of a farming system by giving them a numerical score with an upper limit. Then, weighing and aggregating the information obtained, a score is given to the farm on each of the three scales used to qualify sustainability. The farm with high scores on the scale is considered to be more sustainable (Zahm et al. 2007).

3.2.5 Indicator of Sustainable Agricultural Practice (ISAP)

ISAP was developed by Rigby et al. (2001) as a farm-level indicator of agricultural sustainability for a sample of 80 organic and 157 conventional producers in the United Kingdom. The data used come from a structured questionnaire completed during face-to-face interviews. The information used to generate the ISAP relates to five aspects of horticultural production on the farms: seed source, pest/disease control, weed control, maintenance of soil fertility, and crop management.

The impact of these farming practices on farm sustainability was assessed by identifying criteria commonly adopted for agricultural sustainability from the literature. Then, simple scores were allocated to each farming practice according to whether a particular practice was considered to improve or diminish a farm's performance according to a given criterion. The criteria are discussed below, and the scoring system is then designed (Rigby et al. 2001).

3.2.6 Multiscale Methodological Framework (MMF)

MMF is based on a systems approach from which five general attributes of sustainable natural resource management systems are defined (productivity, stability, resilience, reliability, and adaptability). The attributes are based on scale- and discipline-independent properties. The strategy to derive criteria and indicators from these attributes is part of a general framework for multiscale sustainability evaluation (Lopez-Ridaura 2005).

Operationally, the general framework has a cyclic structure with two phases: (1) a systems analysis phase and (2) a systems synthesis phase. In phase one, sets of criteria and specific indicators for the different scales of analysis are derived. In phase two, results from assessment of the indicators are analyzed, comparing different alternatives through scenario analyses. The results from the evaluation process serve as the basis for the design and implementation of alternatives aiming at greater sustainability, taking into account the objectives of stakeholders at different scales (Lopez-Ridaura 2005).

3.2.7 Response-Inducing Sustainability Evaluation (RISE)

RISE can be used globally for sustainability assessment, through analysis and comparison, of all kinds of farms and production systems. RISE is a management tool that provides an instrument to visualize strengths and potentials (providing a testimonial) but also to identify weaknesses (need for action) regarding the sustainability of the farmer's specific production practices (Häni et al. 2007).

The RISE analysis shows ecological, economic, and social aspects of the sustainability of agricultural production. It uses 12 indicators calculated from more than 60 parameters. The Driving Force-State-Response (DSR) framework is the principle followed in the RISE model. Each indicator contains parameters to outline the State (S) of the system or describe a pressure on or Driving force (D) within the system, driving it in a certain development direction. State parameters have a range of values between 0 (worst case) and 100 (best case). The difference between S and D is the Degree of Sustainability (DS) (Häni et al. 2007).

3.2.8 Sustainability Assessment of Farming and the Environment (SAFE)

The SAFE framework is designed for three spatial levels: the parcel, the farm, and a higher spatial level that can be the landscape, the region, or the state (Van Cauwenbergh et al. 2007).

SAFE is hierarchical and is composed of principles, criteria, indicators, and reference values in a structured way. Principles are related to the multiple functions of the agro-ecosystem, not only the production function. Criteria are specific objectives relating to a state of the system that are easier to assess and to link indicators. Indicators are indicative of the state of the system in an objectively verifiable way, describing features of the agro-ecosystem or elements of prevailing policy, management conditions, and human driving forces. A representative picture of the sustainability of agricultural systems in all its environmental, economic, and social aspects is provided by the set of indicator values. The desired level of sustainability for each indicator, established on a scientific or empirical basis, is the reference value. Indicators and reference values are the operational tools that are used for evaluating the sustainability of the agro-ecosystems. They are the end products of the SAFE framework. This method is used as an assessment tool for the identification, development, and evaluation of farming systems, techniques, and policies, and it is not intended to find a common solution for sustainability in agriculture as a whole (Van Cauwenbergh et al. 2007).

3.2.9 Sustainability Solution Space (SSP) for Decision Making

Wiek and Binder (2005) present an approach to constructing SSP for decision making that gives a concise guideline for sustainable decisions to the decision makers and makes them aware of the synergistic and contradictory effects of their decisions.

The SSP process consists of a prerequisite phase and systemic, normative, and integrative modules. In the prerequisite phase are defined, in a consensus-building process, the goals to be assessed. A sufficient system description is provided in the systemic module by selecting indicators and analyzing interlinkages among them. The second module aims to define sustainability ranges for each of the selected indicators: this is a minimum and maximum value according to the selected criteria and incorporates the scientific judgments, values, and preferences of stakeholders. Finally, the integrative module combines the systemic and normative modules and provides a SSP for the agricultural sector of a defined region. The SSP are within ranges of the values of the indicators that can vary without hampering the sustainability of the whole system. The transdisciplinary process can be envisioned at each step of SSP and ensures that the knowledge and values of the regional stakeholders are included. Also, the transdisciplinary approach improves the soundness of the sustainability assessment and supports the implementation of the elaborated strategies (Binder and Wiek 2007).

References

Binder, C., & Wiek, A. F. (2007). The role of transdisciplinary processes in sustainability assessment of agricultural systems. In F. Häni, L. Pintér, & H. Ferren (Eds.), *From common principles to common practice.* Proceedings and Outputs of the First Symposium of the International Forum on Assessing Sustainability in Agriculture (INFASA) (pp. 33–48). Bern: INFASA and Swiss College of Agriculture.

FAO. (1993). *FESLM: An international framework for evaluating sustainable land management* (World Soil Resources Reports No. 73). Roma: FAO.

Häni, F., Stämpfli, A., Gerber, T., Porsche, H., Thalmann, C., & Studer, C. (2007). RISE: A tool for improving sustainability in agriculture. A case study with tea farms in Southern India. In F. Häni, L. Pintér, & H. Ferren (Eds.), *From common principles to common practice.* Proceedings and Outputs of the First Symposium of the International Forum on Assessing Sustainability in Agriculture (INFASA) (pp. 121–148). Bern: INFASA and Swiss College of Agriculture.

Hansen, J. W. (1996). Is agricultural sustainability a useful concept? *Agricultural Systems, 50*, 117–143.

Lopez-Ridaura, S. (2005). Multiscale methodological framework to derive criteria and indicators for sustainability evaluation of peasant natural resource management system. *Environment, Development and Sustainability, 7*, 51–69.

Marzall, K. (1999). Indicadores de Sustentabilidade para Agroecossistemas. Master dissertation, Faculdade de Agronomia da Universidade Federal do Rio Grande do Sul.

Masera, Ó., Astier, M., & López-Ridaura, S. (2000). El Marco de Evaluación MESMIS. In O. Masera & S. López-Ridaura (Eds.), *Sustentabilidad y Sistemas Campesinos* (pp. 13–44). México: GIRA A. C., Mundi Prensa, and PUMA.

Pervanchon, F. (2007). When farmers integrate sustainable development in their strategy thank a tree: The sustainable farm tree. In F. Häni, L. Pintér, & H. Ferren (Eds.), *From common principles to common practice.* Proceedings and Outputs of the First Symposium of the International Forum on Assessing Sustainability in Agriculture (INFASA) (pp. 111–120). Bern: INFASA and Swiss College of Agriculture.

RAD and CIVAM. (2010). *Diagnostic de durabilité du Réseau Agriculture Durable. Guide de l'utilisateur.* Cesson-Sevigne: RAD and CIVAM.

Rigby, D., Woodhouse, P., Young, T., & Burton, M. (2001). Constructing a farm level indicator of sustainable agricultural practice. *Ecological Economics, 39*, 463–478.

Smith, C., & McDonald, G. (1998). Assessing the sustainability of agriculture at the planning stage. *Journal of Environment Management, 52*, 15–37.

Van Cauwenbergh, N., Biala, K., Bielders, C., Brouckaert, V., Franchois, L., Cidad, V., Hermy, M., Mathijs, E., Muys, B., Reijnders, J., Sauvenier, X., Valckx, J., Vanclooster, M., Van der Veken, B., Wauters, E., & Peeters, A. (2007). SAFE – A hierarchical framework for assessing the sustainability of agricultural systems. *Agriculture, Ecosystems & Environment, 120*, 229–242.

Vilain, L. (2008). *La Méthode IDEA: Indicateurs de Durabilité des Exploitations Agricoles: Guide d'Utilisation.* Dijon: Educagri éditions.

Wiek, A., & Binder, C. (2005). Solution spaces for decision-making: A sustainability assessment tool for city-regions. *Environmental Impact Assessment Review, 25*, 589–608.

Zahm, F., Viaux, P., Girardin, P., Vilain, L., & Mouchet, C. (2007). Farm sustainability assessment using the IDEA method. from the concept of farm sustainability to case studies on French farms. In F. Häni, L. Pintér, & H. Ferren (Eds.), *From common principles to common practice.* Proceedings and Outputs of the First Symposium of the International Forum on Assessing Sustainability in Agriculture (INFASA) (pp. 77–110). Bern: INFASA and Swiss College of Agriculture.

Part II
Multiple Aspects to Develop Methodologies for Sustainability Assessment of Farming Systems

Chapter 4
Normative, Systemic and Procedural Aspects: A Review of Indicator-Based Sustainability Assessments in Agriculture

Claudia R. Binder and Giuseppe Feola

Abstract Methods for assessing the sustainability of agricultural systems do often not fully (i) take into account the multifunctionality of agriculture, (ii) include multidimensionality, (iii) utilize and implement the assessment knowledge and (iv) identify conflicting goals and trade-offs. This chapter reviews seven recently developed multidisciplinary indicator-based assessment methods with respect to their contribution to these shortcomings. All approaches include (1) normative aspects such as goal setting, (2) systemic aspects such as a specification of scale of analysis and (3) a reproducible structure of the approach. The approaches can be categorized into three typologies: first, top-down farm assessments, which focus on field or farm assessment; second, top-down regional assessments, which assess the on-farm and the regional effects; and third, bottom-up, integrated participatory or transdisciplinary approaches, which focus on a regional scale. Our analysis shows that the bottom-up, integrated participatory or transdisciplinary approaches seem to better overcome the four shortcomings mentioned above.

Keywords Sustainability assessment • Indicator • Agriculture • Sustainability solution space

---

C.R. Binder (✉)
Department of Geography, LMU University,
Luisenstraße 37, 80333 Munich, Germany
e-mail: claudia.binder@lmu.de

G. Feola
Department of Geography and Environmental Science, University of Reading,
Whiteknights, RG6 6AB Reading, UK
e-mail: g.feola@reading.ac.uk

A.A. Marta-Costa and E. Silva (eds.), *Methods and Procedures for Building Sustainable Farming Systems: Application in the European Context*,
DOI 10.1007/978-94-007-5003-6_4, © Springer Science+Business Media Dordrecht 2013

4.1 Introduction

Sustainability within agricultural systems is widely discussed and is viewed as essential for the transition towards global sustainable development in international form (UNCED 1992; OECD 2001; WSSD 2002). Despite wide consensus on its relevance, a high degree of variability can be observed both in how sustainable development in agriculture is defined and how it is practically pursued in the policy-making process. The lack of agreement about the definition has brought some researchers (e.g. Hansen 1996) to question the usefulness of the concept of "agricultural sustainability".

The variability existing in the policy-making arena is mirrored and supported by the academic debate, where multiple and sometimes contradictory perspectives coexist on how sustainable development in agriculture should be defined and pursued. Consequently, a wide variety of tools and methods have been developed to assess sustainable development in agriculture, which include among others (i) indicator lists (e.g. Girardin et al. 2000; Rigby et al. 2000; Woodhouse et al. 2000; van der Werf and Petit 2002), (ii) environmental assessment of production alternatives (as in LCA, van der Werf and Petit 2002), (iii) indexes or ecopoints (Taylor et al. 1993; Mayrhofer et al. 1996; van der Werf and Petit 2002), (iv) linear programming models (Rossing et al. 2007) and (v) trade-off models of production alternatives, considering economic, ecological and health aspects (Crissman et al. 1998). The majority of methods developed, however, have focused on ecological aspects and reflect the foci set in sustainable agriculture which is often related to issues such as integrated pest management, organic farming, biodynamic farming, low-input agriculture, agro-ecology, low-input sustainable agriculture and low external input sustainable agriculture (Rigby and Caceres 1997).

There are four main shortcomings in sustainability assessment in agriculture:

1. The multifunctionality in agriculture is often not specifically addressed in sustainability assessments (Rossing et al. 2007).
2. There is an imbalance in the modelling and assessment work performed regarding the three dimensions of sustainability, that is, ecological, economic and social aspects (von Wirén-Lehr 2001), in favour of the ecological one.
3. Research has so far focused on filling important gaps in knowledge and technology, but has omitted to include the step towards utilization and implementation of this knowledge (Rossing et al. 2007).
4. The assessment results themselves are difficult to implement in decision-making, as conflicting goals and the interaction between indicators have not been sufficiently considered (Morse et al. 2001).

As many different approaches exist, which differ in terms of, for example, goal, methods and assessment procedure, different performances are expected, with respect of the four above-mentioned shortcomings. In this chapter, we compare seven indicator-based approaches for sustainability assessment in agriculture in terms of the normative, systemic and procedural dimensions in the assessment procedure (Wiek

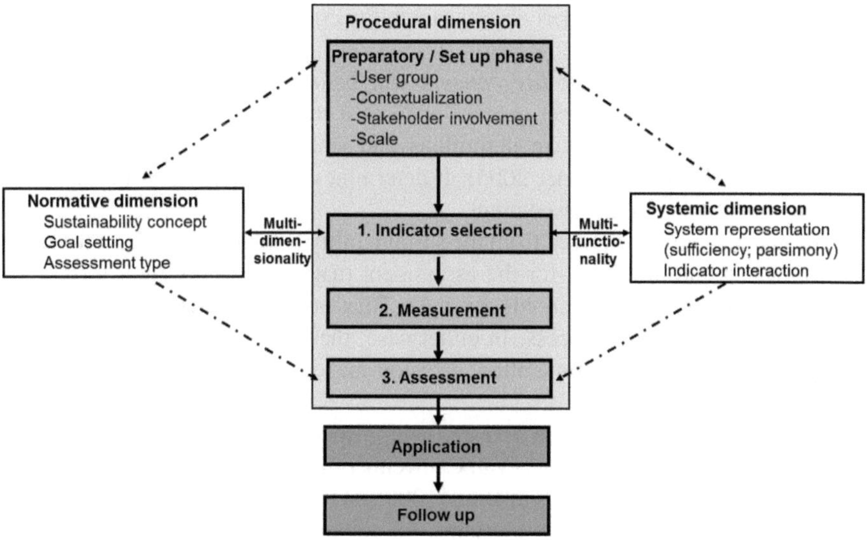

Fig. 4.1 The interrelationship of the normative, systemic and procedural dimensions within the assessment process (After Wiek and Binder 2005; Binder et al. 2010)

and Binder 2005). The analysis and comparison allow for highlighting advantages and disadvantages of the methods and pointing out trade-offs and opportunities for improving the practice of sustainability assessment in agriculture.

4.2 Methodological Approach

Figure 4.1 depicts the assessment process and how the normative, systemic and procedural dimensions are interlinked. In the preparatory phase within the procedural dimension, the user group, the involved stakeholders and their type of involvement (e.g. participatory, transdisciplinary, expert input) are determined. This step, to a large extent, drives the normative and systemic aspects such as the sustainability concept chosen and system representation. In turn, the normative and systemic dimensions affect the preparatory phase, the selection of the indicators and the assessment itself.

4.2.1 Normative Dimension

The consideration of the normative dimension is essential if the indicator-based decision-making system is to be useful for assessment and application. Three issues have to be considered: (i) underlying sustainability concept, (ii) goal setting and (iii) assessment type (Fig. 4.1).

The underlying *sustainability concept* can be completely theory-based (i.e. Niemeijer 2002; Bossel 1999) or developed in a transdisciplinary procedure, in which, for example, legislative definitions and stakeholder perspectives can be included (see Wiek and Binder 2005). It determines to which extent multidimensionality is included in the assessment.

The *goals* should be derived from the sustainability concept. They operationalize the former and are the basis for the assessment that can take different forms, for example, the reference to thresholds or ranges. They can be derived by the researchers or in a transdisciplinary process. In either case, these goals need to be internally consistent and at the same time allow decision-makers flexibility for taking action (Wiek and Binder 2005).

Finally, the indicators can be *assessed* with respect to regulatory standards (e.g. nitrogen in groundwater), targets (Van Cauwenbergh et al. 2007), thresholds (Zahm et al. 2006) and ranges (Wiek and Binder 2005). Of crucial importance is whether the indicators are aggregated into groups, for example, social, economic and ecological, and how the groups are weighted.

It should be considered that normative concepts may vary along cultures and parts of the society (Empacher 2002), and, thus, the question to which extent the assessment is applicable to other countries has to be critically studied before extrapolating results or methodologies to other contexts (Binder and Wiek 2007).

4.2.2 Systemic Dimension

The systemic dimension plays an essential role when selecting and designing the indicators for the assessment. For obtaining an adequate system representation, three issues should be considered: (i) parsimony, (ii) sufficiency and (iii) indicator interaction.

In general, a system should be represented with as much simplicity as possible (parsimony) and as much complexity as necessary (sufficiency). This implies that, for obtaining an adequate system representation, the most relevant relations among the indicators have to be considered in the analysis (Wiek and Binder 2005; Binder and Wiek 2007). The indicators and their relations have to represent the main structures, processes and functions of the economic, ecological and social fields of the system studied and have to refer to the problems and targets to be tackled and thus are linked to the normative dimension.

4.2.3 Procedural Dimension

We structure the procedural dimension into the procedure itself and stakeholder involvement.

4.2.3.1 Structure of the Procedure

As mentioned above, the assessment protocol has to be complete, consistent and replicable if the results should be reproducible, used for benchmarking, to monitor system changes over time or to evaluate the utility of measures taken. We divide the sustainability assessment process into ideal sub-phases. The sequential presentation may not always correspond to the real implementation, which is characterized by feedback loops and cyclical stages. We defined a preparatory phase and five main steps (Fig. 4.1). In the preparatory or set-up phase, the basic elements of the assessment are defined, that is, the system under consideration, the scale of analysis and the user groups of the stakeholders to be involved and the type of their involvement. The core part of the assessment includes three main steps:

1. First, the selection of the indicators is linked to the normative and systemic aspects mentioned above. It should be based on the specific characteristics of the field, farm or region and the problems existing in the selected system. Important criteria for the selection of indicators should be (Binder and Wiek 2001; Scholz and Tietje 2002; Zhen and Routray 2003; Wiek and Binder 2005): (i) goal orientation, (ii) system representation and (iii) data availability. The results of this step include the information on goal specificity of the indicator set (i.e. how well the indicator fits the goals set), its multidimensionality and multifunctionality and the scale of analysis (Smith and McDonald 1998; von Wirén-Lehr 2001; Niemeijer 2002; Payraudeau and van der Werf 2005). In this step, the decision is taken of whether or not to include the interaction of indicators and how it will be implemented.
2. Second, the indicator measurement is related to quantification of the indicators and processes. This can be based on statistical data, surveys or qualitative data.
3. Third, in the assessment, the normative and systemic aspects are included again (Fig. 4.1). Here, one should distinguish between the aggregation and integration of indicators and the specific assessment procedure (Binder et al. 2010).

Then follow the application, and in the final follow-up phase, the results are reported, management advice developed and the indicators monitored over time.

4.2.3.2 Stakeholder Involvement

For an indicator-based sustainability assessment to comprehensively and reliably reflect the salient features of the system, the research and results must be pursued in a society- and policy-conscious framework. We consider participatory and transdisciplinary research methods as essential for doing so (Ravetz 1999; Thompson Klein et al. 2001; Binder and Wiek 2007). It has to be noted that in the assessment process as depicted in Fig. 4.1, the decision when and how to involve stakeholders is already taken in the preparatory phase, indicating this to be a key decision in any procedure.

4.3 Short Overview of the Selected Approaches

Seven approaches were selected because they address the three above-mentioned dimensions: (i) systemic view by providing adequate criteria for system representation, (ii) normative view by including assessment criteria and (iii) procedural component by providing a structure to the assessment. Most of the approaches selected are recently developed approaches, one of which (SSP) has just recently been applied to the agricultural system (Castoldi et al. 2007). One distinction of the selected approaches is the system boundary ranging from focus on farm level to regional scale or across scales (Tables 4.1 and 4.2).

The Indicateur de Durabilité des Exploitacions Agricoles (IDEA) analyzes the sustainability at a farm level addressing several premises. A farm must be able to be

Table 4.1 Overview of the selected approaches: farm level

Approach	Aim	Target group	Definition of sustainable agriculture
IDEA	To provide an operational tool for sustainability assessment at farm level	Planners, policymakers, researchers, farmers, farmer organizations	– Economic viability – Social livability – Environmental reproducibility
ISAP	To operationalize agricultural sustainability in order to support policy making	Researcher and policymakers	– Minimization of off-farm inputs – Minimization of non-renewable resources – Maximization of natural biological processes – Promoting local biodiversity – Enhancing farmers' life quality – Increasing farmers' self reliance – Sustaining farms' profitability – Improving equity – Meeting society's needs for food and fibre
RISE	To provide a simple and cheap but holistic tool to: (1) evaluate the degree of sustainability at farm level and (2) visualize potentials and failures, thus inducing management responses	Farmers	– Productivity – Competitiveness – Efficiency – Protection and improvement of the natural environment and socio-economic conditions of local communities

Table 4.2 Overview of the selected approaches: regional level or across scales

Approach	Aim	Target group	Definition of sustainable agriculture
FESLM	To guide analysis of land use sustainability, through a series of scientifically sound, logical steps. It is integrative (considers all interacting factors), concerned with evaluation, systematic	Planners	Productivity Security Protection Viability Acceptability
MMF	To assess multiscale sustainability with emphasis on peasant agriculture and natural resource management	Researcher and policymakers	Productivity Stability Resilience Reliability Adaptability
SAFE	To identify, develop and evaluate agricultural production systems, techniques and policies	Researcher and policymakers	Biological diversity Productivity Regeneration Capacity Vitality Ability to function
SSP	To identify the Sustainability Solution Space in which stakeholders can find solutions and the system remains or becomes more sustainable	All stakeholders affecting systems' sustainability planners, farmers, policymakers	Theory-based combined with a transdisciplinary process. Includes multidimensionality and multifunctionality

viable in economic terms, livable for the farmer and his family, and ensure the reproducibility of the environment (Zahm et al. 2006). A total set of 41 indicators is derived accounting for these dimensions.

The Indicator of Sustainable Agricultural Practice (ISAP) focuses on the sustainability of specific agricultural practices. The developed index serves in particular "to compare the relative hazards to sustainability posed by different farming methods" (Rigby et al. 2001). It allows for an assessment with limited data availability.

The Response-Inducing Sustainability Evaluation (RISE) (Häni et al. 2003; Porsche et al. 2004) allows for analyzing and comparing the sustainability of a diversity of agricultural production systems or farms. It balances between the straightforwardness of the analysis, the complexity of the reality and the transparency of the results, making so the output comprehensible for a wider public and applicable by farmers.

The Framework for the Evaluation of Sustainable Land Management (FESLM) (Smyth and Dumanski 1993) provides a strategic framework approach for evaluating sustainable land management. It departs from the premise that sustainability is not rigid but has to be capable to capture changes in typologies of areas and development over time. The framework "offers the possibility of providing preliminary estimates of acceptable reliability, without waiting for all of the final data" (Smyth and Dumanski 1993).

The Multiscale Methodological Framework (MMF) (Lopez-Ridaura 2002, 2005) aims at assessing sustainability at multiscale level with emphasis on peasant agriculture and natural resource management. It is based on a discipline-independent systems approach and aims at "building a multi-stakeholder and object driven platform in which objectives and constraints of the stakeholders are coupled to the attributes in order to arrive at useful sets of criteria and specific indicators, meaningful to the stakeholders at different scales" (Lopez-Ridaura 2005).

The Sustainability Assessment of the Farming and the Environment (SAFE) (Van Cauwenbergh et al. 2007) proposes a holistic, hierarchical methodology for assessing the sustainability of the agro-ecological system. SAFE analyzes the effect of farm activities at plot, farm and regional level.

The Sustainability Solution Space for Decision-Making (SSP) (Wiek and Binder 2005; Binder and Wiek 2007; Castoldi et al. 2007; Binder et al. 2008, 2010) is a systemic, multidisciplinary and, as far as possible, a dynamic approach, thanks to the analysis of the links between the indicators used. The method uses indicators' targets in the form of ranges. "A sustainability range of an indicator is the largest range within which a sustainable development can take place" (Wiek and Binder 2005). The result is the largest Sustainability Solution Space possible, which is determined through the examination of consistencies and contrasts between the ranges and through the ranking and composition of targets.

4.4 Results and Discussion

The analysis of the normative, systemic and procedural characteristics of the selected approaches allowed for identifying similarities and differences among the methods. We group the methods in three types: top-down farm assessment, top-down regional assessment, and bottom-up integrated participatory or transdisciplinary assessment. In the following, the typology of the approaches is presented, and the advantages and disadvantages for each group are discussed.

Figure 4.2 illustrates the focus of each method with respect to the normative, systemic and procedural dimension discussed. The methods can be structured in three typologies as follows:

1. *Top-Down Farm Assessment (RISE, IDEA, ISAP).* This group relates to the methods which focus on assessing a farm or a field. The user group is usually the farmer himself or industry working with farmers groups, and no participation occurs. Consequently, the indicators are derived top-down, and the way on how they have to be measured and calculated is determined by a clearly structured methodological procedure. Some of these methods tend to focus on ecological aspects

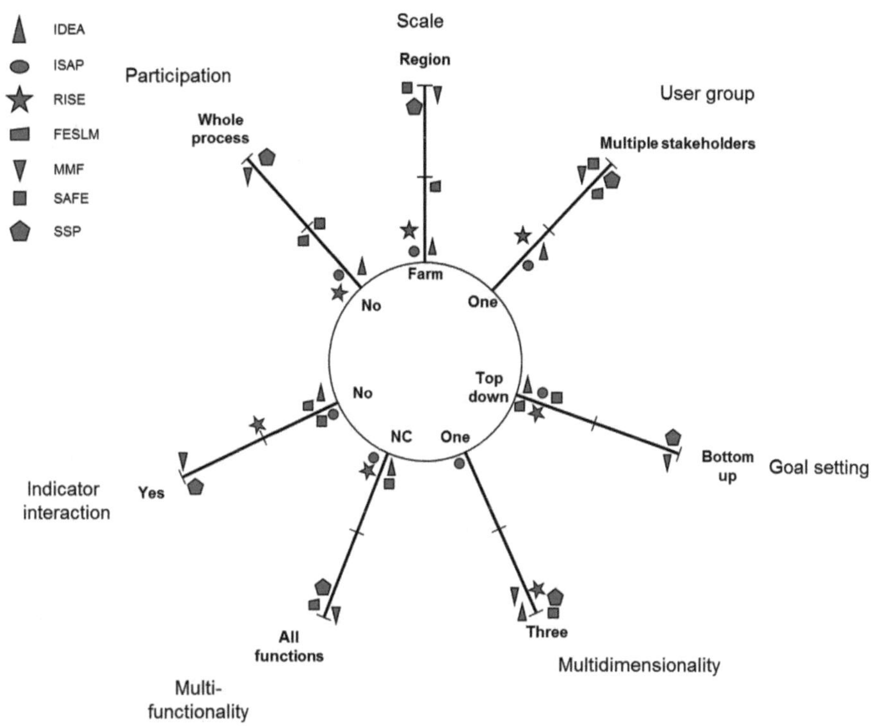

Fig. 4.2 Comparison of the seven approaches with respect to the principal indicator of the normative, systemic and procedural dimension (Binder et al. 2010)

or try to include to some extent also the economic and social perspectives of sustainability but do not consider the multifunctionality of agriculture. Finally, indicators interaction is not taken into account, even though composed indicators are built, for example, in RISE (Häni et al. 2003, 2007). The results from these methods can relatively easily be discussed with farmers, and the procedure allows for monitoring and to some extent benchmarking across regions.

2. *Top-Down Regional Assessment with Some Stakeholder Participation (FESLM, SAFE).* This group relates to methods which study the regional scale or are applicable to the farm as well as the regional level. They include stakeholder participation in the indicator development and have usually multiple stakeholders who are likely to use the results. They always include the ecologic, economic and social dimension of sustainability. However, they do not consider the interrelationship among the indicators, impeding the analysis of trade-offs when designing measures. FESLM translates global concerns to the farm level, whereas SAFE claims to be applicable by both farmers and decision-makers.
3. *Bottom-Up Integrated Participatory or Transdisciplinary Approach (MMF, SSP).* This group refers to methods which ideally focus at the regional scale with multiple stakeholders as user group. They include stakeholder participation throughout the process, including the goal setting process and complement it with theoretical

scientific knowledge (SSP). The system is represented including the interrelationship among the indicators, and the assessment relies on a combination of quantitative (e.g. linear programming) and qualitative (e.g. workshops, expert interviews) tools. The bottom-up process and the large extent of stakeholder involvement support the likeliness that the results will be applied and make the assessment tool flexible for different contexts, yet it makes monitoring and benchmarking across regions extremely difficult.

Concerning multidimensionality, which refers to the normative dimension, the three typologies perform uniformly. That is, the assessment is based on a multidimensional definition of sustainability. Furthermore, it is also uniformly acknowledged that indicators referring to the three dimensions have to be measured separately and not aggregated in a single index. Therefore, the reviewed methods overcome the shortcoming represented by the imbalance of the three sustainability dimensions observed by von Wirén-Lehr (2001) in the practice of sustainability assessment in agriculture.

Concerning indicators interaction and multifunctionality, both referring to the systemic dimension, a significant difference is observed between the top-down (typologies 1 and 2) and the bottom-up (typology 3) approaches. In effect, the methods grouped in the typologies 1 and 2 do not consider either multifunctionality or the interactions among indicators. This represents a disadvantage because these assessment methods may not achieve an adequate system representation. On the other hand, typology 3 considers both multifunctionality and interactions. In this respect, it can be argued that these approaches are able to render a more complex and complete picture of the system's functioning. This is achieved by approaching the procedural dimension in a different way, that is, (i) by involving different stakeholders and especially expert and laymen, (ii) by adapting the indicators' list to the characteristics of each specific system and (iii) by integrating ad hoc developed quantitative (e.g. trade-off analysis, linear programming) with qualitative (e.g. workshops, scenario building and analysis) assessment tools. Stakeholder participation, which in typology 3 is combined with a high adaptability to the specific context under assessment, is likely to enhance the applicability of the results (Ravetz 1999; Binder and Wiek 2007), thus supporting to meet the need expressed by Rossing et al. (2007) of bridging knowledge and implementation of the knowledge. Interestingly, the applicability in one system is achieved at the expenses of the reproducibility and benchmarking of the results among different systems, as the assessment (i.e. indicators selection, assessment goals and criteria) is extremely tailored to the specific system under assessment. Furthermore, due to the participation of different stakeholders, the need to select the indicators and to define the scale of analysis and the border of the system, the assessment procedure may tend to be time- and resource-consuming, which represents an obvious disadvantage. Such a characterization in terms of applicability of the methods grouped in typology 3 is significantly different to that of methods grouped in typology 1. The latter are characterized by a relatively "easy" procedure, which is highly standardized and reproducible (e.g. preselected indicators, system definition and scale of analysis), which also allows for benchmarking and comparison among different systems. However, the absence of stakeholder participation and the low adaptability of the assessment procedure and tools to the specific system are likely to reduce the applicability of the assessment results.

The methods grouped in typology 2 show similarities, in terms of applicability, with both typologies 1 and 3. For example, stakeholder participation is considered an option but is not structurally integrated in the assessment procedure. Similarly, indications concerning the indicators to be used exist, but there is no predefined selection to be adopted as standard in different contexts. Because this typology is characterized by leaving a significant room for the researcher in orienting the assessment's procedure, it may show a mixture of the advantages and disadvantages, which distinguish typologies 1 and 3.

In summary, all the typologies are characterized by strength and weaknesses. However, from an overall perspective, the methods grouped in the typology 3 seem to better overcome the four shortcomings of sustainability assessment in agriculture mentioned above. They are multidimensional, multifunctional and explicitly consider interactions among the indicators. Furthermore, they strongly address the applicability of the results by involving the stakeholders in the assessment procedure and providing them scenarios (MMF) or a space for decision-making (SSP) which can support them in sustainably developing their system.

4.5 Conclusions

This chapter provided a review of seven indicator-based assessment approaches for agriculture. These approaches were analyzed with respect to three dimensions: a normative, a systemic and a procedural one. Such an analysis shows how these approaches only partially fulfil the current needs on agricultural sustainability assessment, namely, (i) multifunctionality of agriculture, (ii) multidimensionality (balance between ecological, economic and social aspects), (iii) create base for making a step towards utilization and implementation of the assessment knowledge and (iv) identify conflicting goals and trade-offs by including the interaction between indicators. This chapter highlighted the advantages and disadvantages in the way the steps of the assessment are pursued, that is, goal setting, choice of assessment type, indicators' selection and aggregation or integration, structure of the procedure and stakeholders' involvement. In doing so, three types of indicator-based assessments were identified: (i) top-down farm assessment; (ii) top-down regional assessment with some stakeholder participation; (iii) bottom-up regional approaches with participation throughout the assessment process; and (iv) transdisciplinary integrated assessment. Each of these assessment types has specific advantages and disadvantages. If, however, the four above-mentioned shortcomings are to be overcome, the authors recommend to performing a transdisciplinary integrated assessment. The method proposed for doing so is the Sustainability Solution Space (SSP). The approach allows for obtaining a Sustainability Solution Space within which stakeholders and policymakers can take their decisions, knowing that they are still within a sustainable path. The space is constructed by utilizing on the interaction between indicators, which furthermore provides the basis for a trade-off analysis when assessing strategies for improving the sustainability of the system. Finally, stakeholder involvement occurs in different phases, allowing for ownership of the results and a higher probability of their implementation.

Acknowledgements The authors thank Christina Kühnhauser for formatting this chapter. An earlier and more extensive version of this chapter was published in Environmental Impact Review (Binder C. R., Feola G., Steinberger J. (2010), Considering the normative, systemic and procedural dimensions in indicator-based sustainability assessments in agriculture, *Environmental Impact Assessment Review, 30*, 71–81).

References

Binder, C., & Wiek, A. (2001, November 12–14). *Sustainability spaces: A new concept to evaluate development using indicator systems.* Paper presented at Inaugural Conference for Industrial Ecology, Netherlands.

Binder, C. R., & Wiek, A. (2007). The role of transdisciplinary processes in sustainability assessment of agricultural systems. In F. J. Häni, L. Pintér, H. R. Herren (Eds.), *From common principles to common practice* (pp. 33–48). Proceedings and Outputs of the first Symposium of the International Forum on Assessing Sustainability in Agriculture (INFASA). Bern: International Institute of Sustainable Development and Swiss College of Agriculture.

Binder, C. R., Steinberger, J., Brundiers, K., Schmidt, H., Schmid, A. (2008). *Nachhaltigkeit in der Schweizerischen Milchwertschöpfungskette*. Schlussbericht, Soziale und Industrielle Ökologie, Geographisches Institut. Zürich: Universität Zürich, 35 p.

Binder, C. R., Feola, G., & Steinberger, J. (2010). Considering the normative, systemic and procedural dimensions in indicator-based sustainability assessments in agriculture. *Environmental Impact Assessment Review, 30*, 71–81.

Bossel, H. (1999). *Indicators for sustainable development: Theory, methods, applications.* Winnipeg: International Institute for Sustainable Development.

Castoldi, N. B. L., Steinberger, J., Binder, C. R. (2007). Sustainability solution space using agro–ecological indicators at field level. In M. Donatelli, E. A. (Eds.), *Farming systems design 2007* (pp. 99–100). International Symposium on Methodologies on Integrated Analysis on Farm Production Systems, Catania.

Crissman, C. C., Antle, J. M., & Capalbo, S. (Eds.). (1998). *Economic, environmental and health tradeoffs in agriculture: Pesticides and the sustainability of Andean potato production.* Dordrecht: Kluwer Academic.

Empacher, C. (2002). *Soziale Dimensionen Der Nachhaltigkeit. Perspectiven der Konkretisierung und Operationalisierung*. Frankfurt: ISOE.

Girardin, P., Bockstaller, C., & Van der Werf, H. (2000). Assessment of potential impacts of agricultural practices on the Environment: The AGRO*ECO method. *Environmental Impact Assessment Review, 20*, 227–239.

Häni, F., Braga, F., Stämpfli, A., Keller, T., Fischer, M., Porsche, H., et al. (2003). RISE, a tool for holistic sustainability assessment at the farm level. *International Food and Agribusiness Management Review, 6*, 78–90.

Häni, F., Gerber, T., Stämpfli, A., Porsche, H., Thalmann, C., Studer, C., et al. (2007). RISE: A tol for improving sustainability in agriculture. A case study with tea farmers in southern India. In F. J. Häni, L. Pintér, H. R. Herren (Eds.), *From common principles to common practice* (pp. 121–148). Proceedings and Outputs of the first Symposium of the International Forum on

Assessing Sustainability in Agriculture (INFASA). Bern: International Institute of Sustainable Development and Swiss College of Agriculture.

Hansen, J. W. (1996). Is agricultural sustainability a useful concept? *Agricultural Systems, 50*, 117–143.

Lopez-Ridaura, S. (2002). Evaluating the sustainability of complex socio-environmental systems. The MESMIS framework. *Ecological Indicator, 2*, 135–148.

Lopez-Ridaura, S. (2005). Multiscale methodological framework to derive criteria and indicators for sustainability evaluation of peasant natural resource management system. *Environment, Development and Sustainability, 7*, 51–69.

Mayrhofer, P., Steiner, C., Gärber, E., Gruber, E. (1996). *Regionalprogramm Ökopunkte Niederösterreich*. Informationsheft. Wien: NÖ Landschaftsfonds.

Morse, S., McNamara, N., Acholo, M., & Okwoli, B. (2001). Sustainability indicators: The problem of integration. *Sustainable Development, 9*, 1–15.

Niemeijer, D. (2002). Developing indicators for environmental policy: Data driven and theory-driven approaches examined by example. *Environmental Science and Policy, 5*, 91–103.

OECD (Organization for Economic Cooperation and Development). (2001). *Environmental indicators for agriculture. Methods and results*. Paris: OECD Publication Service.

Payraudeau, S., & van der Werf, H. M. G. (2005). Environmental impact assessment for a farming region: A review of methods. *Agriculture, Ecosystems & Environment, 107*, 1–19.

Porsche, H., Fischer, M., Braga, F., Häni F., et al. (2004). *Introduction of the sustainability assessment tool RISE into Canadian agriculture* (Working paper), University of Guelph, Canada, 11 p.

Ravetz, J. (1999). Citizen participation for integrated assessment: New pathways in complex systems. *International Journal of Environment and Pollution, 11*, 331–350.

Rigby, D., & Caceres, D. (1997). *The sustainability of agricultural systems* (Working Paper No. 10). Manchester: Institute for Development Policy and Management, University of Manchester.

Rigby, D., Howlett, D., Woodhouse, P. (2000). *A review of agricultural and rural livelihood sustainability* (Working Paper No. 1, Working Paper No. 2). Department for International Development Research Project No. R7076CA.

Rigby, D., Woodhouse, P., Young, T., & Burton, M. (2001). Constructing a farm level indicator of sustainable agricultural practice. *Ecological Economics, 39*, 463–478.

Rossing, W. A. H., Zander, P., Josien, E., Groot, J. C. J., Meyer, B. C., & Knierim, A. (2007). Integrative modeling approaches for analysis of impact of multifunctional agriculture: A review for France, Germany and the Netherlands. *Agriculture, Ecosystems & Environment, 120*, 41–57.

Scholz, R. W., & Tietje, O. (2002). *Embedded case study methods. Integrating quantitative and qualitative knowledge*. Thousand Oaks/London: Sage.

Smith, C. S., & McDonald, G. T. (1998). Assessing the sustainability of agriculture at the planning stage. *Journal of Environmental Management, 52*, 15–37.

Smyth, A. J., & Dumanski, J. (1993). *FESLM: An international framework for evaluating sustainable land management*. Geneva: Food and Agriculture Organization of the United Nations (FAO).

Taylor, D. C., Mohamed, Z. A., Shamsudin, M. N., Mohayidin, M. G., & Chiew, E. F. C. (1993). Creating a farmer sustainability index: A Malaysian case study. *American Journal of Alternative Agriculture, 8*, 175–184.

Thompson Klein, J., Grossenbacher-Mansuy, W., Häberli, R., Bill, A., Scholz, R. W., & Welti, M. (Eds.). (2001). *Transdisciplinarity: Joint problem solving among science, technology and society: An effective way for managing complexity*. Basel: Birkhäuser.

UNCED (United Nation Conference on Environment and Development). (1992). *Agenda 21 – An action plan for the next century*. New York: Endorsed at the United Nations Conference on Environment and Development.

Van Cauwenbergh, N., Biala, K., Bielders, C., Brouckaert, V., Franchois, L., Garcia Cidad, V., Hermy, M., Mathijs, E., Muys, B., Reijnders, J., Sauvenier, X., Valckx, J., Vanclooster, J., Van der veken, B., Wauters, E., & Peeters, A. (2007). SAFE – A hierarchical framework for assessing the sustainability of agricultural systems. *Agriculture, Ecosystems & Environment, 120*, 229–242.

Van der Werf, H. M. G., & Petit, J. (2002). Evaluation of the environmental impact of agriculture at the farm level: A comparison and analysis of 12 indicator-based methods. *Agriculture Ecosystems & Environment, 93*, 131–145.

Von Wirén-Lehr, S. (2001). Sustainability in agriculture-an evaluation of principal goal-oriented concepts to close the gap between theory and practice. *Agriculture, Ecosystems & Environment, 84*, 115–129.

Wiek, A., & Binder, C. (2005). Solution spaces for decision-making – A sustainability assessment tool for city-regions. *Environmental Impact Assessment Review, 25*, 589–608.

Woodhouse, P., Howlett, D., Rigby, D. (2000). *A framework for research on sustainability indicators for agriculture and rural livelihoods* (Working Paper No. 2). Department for International Development Research Project No. R7076CA.

WSSD (World Summit on Sustainable Development). (2002). *Report of the world summit on sustainable development*. Johannesburg: WSSD (World Summit on Sustainable Development).

Zahm, F., Viaux, P., Girardin, P., Vilain, L., Mouchet, C. (2006). Farm sustainability assessment using the IDEA method: From the concept of farm sustainability to the case studies on French farms. In F. J. Häni, L. Pintér, H. R. Herren (Eds.), *From common principles to common practice* (pp. 77–110). Proceedings and Outputs of the first Symposium of the International Forum on Assessing Sustainability in Agriculture (INFASA). Bern: International Institute of Sustainable Development and Swiss College of Agriculture.

Zhen, L., & Routray, J. L. (2003). Operational indicators for measuring agricultural sustainability in developing countries. *Environmental Management, 32*, 34–46.

Chapter 5
Assessing the Sustainability of Activity Systems to Support Households' Farming Projects

Médulline Terrier, Pierre Gasselin, and Joseph Le Blanc

Abstract This chapter aims to introduce the setting up of an evaluation tool assessing the sustainability of activity systems and supporting farming households' projects at the establishment stage. This chapter analyses three methods used to appreciate the farm sustainability and identifies not only their limits but also their contributions to our own methodology, at the level of complex activity systems in which farming production is combined with transformation, sales or outside activities. We propose to recognise two different contributions to sustainable agriculture: a farm-focused sustainability and an extended sustainability, which means a contribution to the sustainable development at a regional scale. These theoretical elements were regularly confronted with the analysis of advisors' practices and comprehensive surveys with households in Southern France, where an analysis was carried through a partnership with researchers and local actors. It produced a tool to appraise agricultural projects, with pluriactivity or without, distinguishing farm-focused and extended sustainability.

Keywords Sustainability assessment • Farm sustainability assessment methods • Farming households' projects • Pluriactivity • Organisation scales

M. Terrier
CEMAGREF, UR DTM, 2 rue de la papeterie BP 76,
F-38402 Saint Martin d'Hères Cedex, France
e-mail: medulline.terrier@irstea.fr

P. Gasselin (✉)
INRA, UMR951 Innovation,
2 place Viala, F-34060 Montpellier Cedex 1, France
e-mail: gasselin@supagro.inra.fr

J. Le Blanc
ADEAR-LR, Mas Saporta, F-34875 Lattes, France
e-mail: adearlr@yahoo.fr

A.A. Marta-Costa and E. Silva (eds.), *Methods and Procedures for Building Sustainable Farming Systems: Application in the European Context*,
DOI 10.1007/978-94-007-5003-6_5, © Springer Science+Business Media Dordrecht 2013

5.1 Introduction

This chapter aims to introduce the setting up of an evaluation tool assessing the sustainability of activity systems and supporting farming households' projects at the establishment stage.

French authorities have considered farm establishment as a priority for more than 10 years. Farm establishment is crucial to maintaining and developing rural areas. However, farm establishments are not enough to renew the agricultural population with one departure on two not replaced (MAAP 2007). French state supports farm establishment as part of a plan associated with financial support according to eligibility criteria. In 2004, only one third of the farm establishments has benefited from this support (Lefebvre et al. 2006). In Southern France, the regional council of Languedoc-Roussillon proposes its own help facility in the farm establishment support plan and targets farm establishments that are excluded of state support. Concerning this issue, advice structures shape and coordinate farm establishments at the territory level. Financing and advice access depends on standards which are common or specific of these structures inserted in various support frameworks. These standards define, often by an implicit way, what is a sustainable farming project.

The pluriactive farmers represent 20% of farmers and one out of three farming households is pluriactive (Rattin 2002). They often do not benefit from the national farm establishment support (Laurent and Mundler 2006). In France, the agriculture professionalisation trend has marginalised pluriactivity whereas it may be an alternative to the main productive stream. Indeed, it is a residual social form which has demonstrated a strong capacity to resist sector-based and territorial crisis. Because of its resilience, pluriactivity appears to be a pertinent situation upon which to base a sustainability assessment tool.

Talking about sustainable agriculture leads to recognise different agricultural functions: productive and marketable but also environmental and social ones. It is fundamental to empower the extension actors with capacities and tools which enable them to promote sustainable farming projects, whether pluriactive or not. Moreover, these tools should allow project initiators to analyse their project sustainability from a dynamic point of view. Therefore, our purpose is to produce an intermediate object, a mean to support and generate dialogue and learning. This work lead in Aude in the south of France was carried out as part of the action research project Intersama ("Insertion territoriale des systèmes d'activités des ménages agricoles" in Languedoc-Roussillon) in partnership with researchers and actors within the framework of the PSDR3 programme ("Pour et Sur le Développement Régional"). This research work fits with the regional recognition of both farm establishment without national aids and pluriactivity.

The first part of this chapter introduces the conceptual framework we adopt to analyse the farming project sustainability. Then, the two-part method adopted to design the tool is introduced and justified. The third part of this chapter details the results of the analysis and comparison of three sustainability assessment methods and introduces the designed frame tool. The last section proposes improvement perspectives.

5.2 Pluriactive Farming Project Sustainability as Study Focus

The French rural code defines agricultural pluriactivity at the individual scale as the exercise of one or several profitable activities aside from the farm work. This means that all the activities which extend the production act, such as processing or marketing activities, are considered by the rural code as farm activities. However, some authors deem that these activities imply specific abilities and mobilise different networks, so it is consistent to analyse them in terms of pluriactivity (Blanchemanche 2000).

The household seems to be the relevant social entity to examine the farming project sustainability in the economical, social and historical study context. Indeed, a household is not only often a decision-making and managing unit but it is also a residential, consumption and accumulation unit.

Pluriactive household studies require an analysis framework that enables the understanding of relationships between the household, its resources and activities and their consequences on the whole global functioning. To understand pluriactive household production system logic, they should be considered as part of an inclusive system – the activity system – which allows to grasp the interactions between the different activities implemented (Paul et al. 1994). These interactions are various: risk management, activity signification, work organisation, incomes, etc. The household makes activity and resource allocation choices that depend not only on economical but also identity, affective and axiological rationalities. These four rationality registers are expressed in synergy or tension and build the base of the farmer and family decisions coherence. Household farm activities must be set back in a broader activity system without hypothecating on the different activity roles in the system (Mundler et al. 2007). Activities are linked to each other within the activity system by functional and/or temporal and spatial links, and each one plays a specific part in the global dynamics of the system. This dynamic balance may be a durability determinant.

Sustainable agriculture is a sustainable development sector-based declension of the concept with the same definition pitfalls (Landais 1998). Defining what sustainable development means implies the need to specify goals and action standards shared by everybody. Consequently, any stake or action relative to sustainable development should be foreseen considering the different stakeholders' positions and representations. There are varying definitions of sustainable agriculture in the scientific literature. Within a pragmatic perspective, sustainable agriculture means agriculture able to carry out its crop and livestock system reproduction and therefore the natural resources on which they depend. In this way, an "enlightened productivist agriculture" is possible (Deffontaine 2001). Others define sustainable agriculture as an economically viable, ecologically safe and socially fair agriculture (Vilain 2008). It is an "agro-environmental" agriculture which regards environment as a production goal while at the same time taking local actors into consideration, social links maintain and inputs savings. From a more general perspective, sustainable agriculture must satisfy two goals at the same time (Godard and Hubert 2002): (1) be sustainable by and for itself through the use of sustainable practices, referred to here as farm-focused sustainability, and (2) contribute to the sustainable development at a regional scale, in

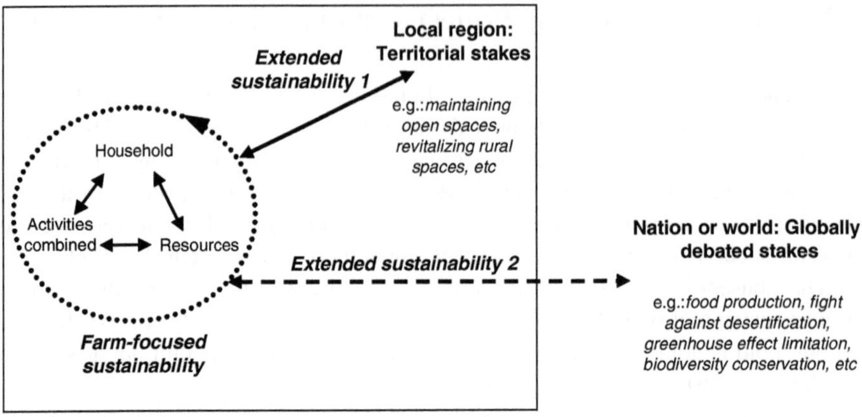

Fig. 5.1 Activity system sustainability refers to different organisation scales

which case we are talking about extended sustainability. Farm-focused and extended sustainability are each distinguishable by the analysis scale at which they must be considered (Fig. 5.1). A farm-focused sustainability assessing scale is the farm or the activity system. This notion is similar to the concept of durability which designates the capacity of the system to maintain itself but is economically and socially limited. Farm-focused sustainability also includes environmental aspects. Social reflections in relation to sustainable development are then referred to the extended sustainability scale. Extended sustainability is the farm contribution to the sustainable development at a regional scale which implies a concrete model definition translated into common goals or at least territorially identified and concerted stakes. Thus, activity system sustainability refers to different organisation levels in relation to stakes of different natures.

This theoretical position determines the configuration of the tool we propose to design. Compared to the current farm creation advisory tools, several points make it original:

1. The activity system concept implies a holistic and not longer farm-focused approach of farming projects, whether pluriactive or not. It differs from the main view promoted by the national plan supporting the young farmer establishments which induces farming project-focused advice. Indeed, project analysis – as it is practised by the advisory structures part in this plan – overlooks the activities or incomes of the other family members or even of the pluriactivity project initiator.
2. The environmental and territorial dimensions recognised in our extended sustainability definition suggest the tool should be contextualised and calibrated in relation to territorial stakes and specificities. In addition, they imply taking a deeper look at the projects, beyond socio-economical dimensions.
3. We propose an *ex ante* activity system sustainability assessment tool whereas farm sustainability assessment tools are commonly *ex post* (Peschard et al. 2004).

4. The tool is designed as guidance support (Paul 2004). It is an intermediate object that supports the relationship between the advisor and the project initiator. It encourages learning and recognition of flexibility and allows to specify the progress priorities. It is not designed as a certification tool determining financing access (Gafsi et al. 2006). It is also non-normative since it does not create any quantitative standards and does not propose a new project sustainability scoring.

5.3 Method

The tool design was led in two different stages. In the first step, we compared three *ex post* farm sustainability assessment methods to define a first group of indicators and examined the various assessment and scoring modalities. Then, this first version tool was tested in the field in order to hone and enhance the indicators and their rules of use.

5.3.1 Comparison of Three Existing Methods

There are many *ex post* farm sustainability assessment tools. We design the first version of our tool from a critical analysis of existing tools. These tools differ from each other by the assessment goal, the analysis and assessment scales (the plot or the farm), the farm productions appraised, the collected data nature, the indicator types (pressure or state, simple or incorporated), the scoring scales and the standard values (Peschard et al. 2004). Each farm sustainability assessment method defines in an implicit way a farm model family of which would be more sustainable than others. Therefore, choosing one of them would constitute judging the underlying model sustainability. Thus, we recognise that sustainability criterions are based on one hand on the individual representations of the designers, and on the other hand, they depend on a social construction related to stakes that are relevant in a specific territorial context. That is why it was impossible to choose between these tools. However, the different sustainability assessment tools are built on several consensuses we attempted to extract through the comparative analysis of three of them (Table 5.1): IDEA (Indicateurs de Durabilité de l'Exploitation Agricole), ARBRE (Arbre de l'Exploitation Aagricole Durable) and RAD (Réseau de l'Agriculture Durable).

These three tools share the common trait of proposing a global farm assessment without focusing on a particular dimension or farm production. They have an educational goal. These shared sustainability indicators constitute our tool frame. A critical analysis of the three methods enables us to select the consensual indicators rank according to the three sustainable development dimensions. The hold indicators had been adapted to our study subject, which is the farming activity system. This set of consensual indicators was structured in a first version of the tool, which was then used as a base to subject this first version to real situations by surveys.

Table 5.1 Characteristics of the three assessment tools

	ARBRE de l'exploitation agricole durable (ARBRE)	Indicateurs de Durabilité de l'Exploitation Agricole (IDEA)	Méthode du Réseau Agriculture Durable (RAD)
Dimensions of sustainability assessed	Transmissibility, reproducibility, viability, liveability	Economical, agro-environmental, socio-territorial	Economical, environmental, social
Agricultural productions assessed	Any agricultural production	Mix crop and livestock productions. Some crops cannot be well assessed by IDEA (market gardening, honey production, small fruits, etc.)	Dairy production
Territory	France	France	Western France
Assessment scale	The farm	The farm	The farm
Indicators	48 Themes declined in 82 indicators	17 Objectives declined in 47 indicators	22 Indicators
Presentation of the results	A tree presents the results: each indicator is represented by a leaf which is coloured or not depending on the assessment results. ARBRE does not suggest any data aggregation, the results are only visual	For each sustainability dimension, a star diagram presents the results. The final score is the lowest of the three scores	For each sustainability dimension, a star diagram presents the results
Assessment type	Qualitative assessment	Quantitative assessment	Quantitative assessment
	Individual assessment lead in a farmer group. For each indicator, the response is judged in comparison to the group average for this indicator and debated in relation to the farmer goal	Individual assessment based on data collected by surveys. For each indicator, a notation system promotes practices in relation to their sustainability	Individual assessment based on data collected by surveys. For each indicator, a notation system promotes practices in relation to their sustainability

Sources: Pervanchon (2004); RAD and CIVAM (2001)

5.3.2 Field Surveys

Fifteen comprehensive surveys with households who combine several activities tested and widely supported this first tool version. Surveys were taken with a sample of households carrying out a farming activity for 3–6 years. The sample gathers various situations with two criteria: (1) the farm establishment path, meeting households having received various advisory and aid plans and (2) the household and farmer activities combined. Following Curie et al.'s advices (Curie et al. 1990), we organise the surveys to cross the three spheres of functional and structural coherence of the activity system, which means the working life (farming and other activities), the private life and the family life. Our activity system aims at identifying strong and weak points as regards to sustainability. To identify these strong and weak points, we analysed together the system functioning and its path by tackling the following points: (1) the economical activities of each household member, (2) the private life, (3) the social life seen through the social networks they belong to and (4) the family and domestic life analysed through the activity system history. The strong and weak points identified, thanks to the analysis, were then confronted to the consensual indicators of the first version of the tool described above. This systematic confrontation enables an iterative and critical enrichment of the tool. At the end of this fieldwork, a last tool confrontation to the theoretical frame allows to complete the assessment tool.

5.4 Results

5.4.1 Analysis of Three Ex Post Evaluation Methods

The three evaluation methods studied (IDEA, ARBRE and RAD) maintain ambiguity on the level on which the indicators and the scales of analysis and evaluation refer: assessing the agricultural activity sustainability implies to estimate its contribution to sustainable development of wider and encompassing organisation levels (territory, nation). Indeed, Allaire and Dupeuble (Allaire and Dupeuble 2004) notice that the individual farming activities use collective resources which are the product of multiple interdependences.

Thus, an elemental aggregation of elements of sustainability at the farming level is not necessarily correlated to proportional effects at the territory level. The sustainability of each part does not guarantee the sustainability of the whole. Some of the indicators really come to the fore at one precise level. Moreover, there are interdependences between farms and the territories, such as the example of the hedges, proposed by Allaire and Dupeuble (*ibid.*). The three methods consider the length of hedges as an indicator of the contribution to landscape protection and biodiversity conservation. Nevertheless, hedges' contribution to landscape protection and biodiversity conservation is more than a simple addition of lengths and requires the contribution and coordination of several actors.

Besides, the three methods studied use, without any discrimination, indicators of farm-focused and extended sustainability. For example, the "quality of life" indicator is about the farm-focused sustainability, whereas the "transmissibility" indicator is about the extended sustainability. According to the people we met during the surveys, "quality of life" is a transversal notion which depends on the relationships to the work, the farm place, the social connection and the welfare but also to the representations of geographical, cultural, professional or affective isolation: the "quality of life" indicator informs about the capacity of the system to maintain itself. In opposition, the "transmissibility of the farm" criterion is linked to the stake of renewing activity and population: it is a criterion of sustainability on the long term at the territory scale.

In agreement with this, we identified for each criterion whether it was belonging to farm-focused or extended sustainability criterion which implies important consequences on the evaluation method. On one hand, farm-focused sustainability is evaluated by indicators informing us about the system capacity to last in the time. Such indicators are identified at the scale of the activities combination by observations and surveys with the households. On the other hand, extended sustainability can only be read through wider knowledge and information (e.g. the environmental sustainability of agricultural practices), in relation to socially shared goals that might be translated to agriculture. Assessing the contribution of the combination of activities to an extended sustainability asks many questions about the spatial and physical analysis scale. How should those goals be defined? Much has been done about the method and position to identify the challenges and representations of the extended sustainability at the territory scale (Chia et al. 2009). Therefore, we characterised those issues in our study combining literature and surveys in the territory.

Another limit of the three evaluation methods lies in their specificities regarding the productions as much as the activity systems and the scoring method. The three methods differ in their final representation of what is a sustainable farm but also in their standards and scoring. When an indicator is traduced in a score, it is based on a scientific reality but also on choices of the designer. So, the scoring scale cannot be dissociated of the ecological and socio-economical context in which the tool was designed. This limits the application field of each method. For example, irrigated maize crops always receive bad score for its water consumption, but it would be a nonsense to penalise the irrigated rice in a French farm of Rhone Delta where water overflows.

Our tool aims to evaluate any type of agricultural production or even any activity in general. Therefore, it is not relevant to calibrate it on technico-economical references that are specific to a production context. Moreover, the three methods analysed require large time survey (1–2 days) and accurate data, and such a precision may not be possible to apply in an *ex ante* method, when the system is not implemented yet. So we have decided to elaborate a qualitative tool that enables us to get free from the problem of threshold and to reduce the quantity of data to be collected. The choice of qualitative evaluation makes easier the consensus about the themes of sustainability to be mobilised.

Besides, we had to deal with the problem of aggregation of the criteria of various activities. The three methods studied evaluate the sustainability at the scale of the farm and do not consider the other activities of the household, except the IDEA methods which merges those activities in one single indicator. The social and economical criteria of various activities can be grouped without methodological problems. But it is not the case for the environmental indicators. As a matter of fact, considering the environmental impact of the combination of activities implies to evaluate and compare very different and remote activities, such as road transport and extensive ovine breeding. We made the choice to evaluate the social and economical sustainability at the scale of the combination of activities but to evaluate the environmental sustainability at the only farm scale.

5.4.2 Modalities of Evaluation

We chose a qualitative evaluation that consists in judging the answer to various themes with regard to the objectives expressed by the person and to the application in his practices. This is not a scoring system. It enables the advisor to evaluate the objective sustainability of the system and at the same time to construct a reflection with the project initiator about the progress of the project. The discussion is about the themes, the indicators that make sense for the project initiator (the farmer, the household) regarding the sustainability of his activities in the territory. The list of indicators can be enriched with new indicators that were not proposed at the beginning. Actually, taking an interest in the goals of the person leads to interrogate his demands and his reflections about sustainability. The analysis done by the advisor and the project initiator leads to identify strengths and weaknesses of the project and to define possibilities of progress.

Regarding the agro-environmental indicators, we refer to the good practices as defined and well marked in the methods ARBRE and IDEA. However, the thresholds for some of these indicators are not relevant in the context of our study. We can take the example of nitrogen fertilisation, a criterion that the three methods use. IDEA and RAD use the indicator of the apparent nitrogen balance, expressed in N kg per hectare. A balance inferior to 30 kg/ha (IDEA) or 20 kg/ha (RAD) gives the best mark; over those thresholds, the excess of nitrogen is penalised. This indicator raises two problems for our tool: (1) as the tool is going to be used for an ex ante evaluation, precise enough fertilisation data do not exist. Even when the project initiator already owns his or her farm, which is not always the case, he or she still has no idea about the precise technical itineraries; (2) the Mediterranean context where we have designed our tool is characterised by chronic deficiency in nitrogen: in this case, what should be considered a factor of unsustainability is the lack of nitrogen and not the excess.

Therefore, we propose that the environmental sustainability indicators do not give a mark or an evaluation but would be a support for discussion, with the aim of striking up a discussion with people who are not stemming from rural nor agriculture. This evaluation method, though debatable, would constitute a significant progress as it introduces environmental concerns in the field of project evaluation in agriculture, a field where it is not currently considered.

Table 5.2 The tool's themes and indicators

Dimension	Themes	Indicators	Emerging indicators
Socio-territorial	14	42	22
Economical	9	19	5
Agro-environmental	9	–	–

5.4.3 Structural Analysis of the Tool

The tool is presented in a table. The table is divided into the three axes of sustainable development and agriculture: socio-territorial, economical and agro-environmental. Each axis is divided into themes and each theme into indicators (Table 5.2). All the factors of sustainability we have identified in the surveys enter in this table. Some of these did not appear in the first version of the table we had built from bibliography, for example, the n. A10 factor: "distribution of the tasks between the household members".

The structure of the tool represents the differentiated activity system contributions to both farm-focused and extended sustainabilities. When the indicator assesses a system contribution to the sustainable development of its territory, the table mentions to which stake it refers. For example, the "animal and vegetal biodiversity" indicator contributes to three environmental concerns: renewing biodiversity, breaking up risks and protection landscape. So each indicator contributes to estimate strength and weaknesses of the project (Table 5.3).

As we told it before, the tool evaluates the farm-focused sustainability at the scale of the activity system for the social and economical indicators but reduces the environmental factors to the agricultural activity only. Mundler (Mundler 2009) perfects the concept of durability (in the sense of "lasting") of the activity systems distinguishing two pillars of resources (internal and external) that we propose to mobilise for the evaluation of the farm-focused sustainability. Thus, the farm-focused sustainability lies in (1) an internal farm-focused sustainability originated in the members of the households, their resources (being the economical and social capital some internal resources), their activities and the interaction between these activities through the knowledge of the household and (2) an external farm-focused sustainability due to the territory where the activities take place, since several resources depend on this territory: institutional context of the farms like the national and community agricultural policies, rules and standards, other actors' logics and territorial logics.

In order to traduce this dichotomy, we propose to analyse the farm-focused sustainability using the SWOT method (for strengths, weaknesses, opportunities and threats). We separate on one hand the strengths and weaknesses that are internal to the activity system and on the other hand the opportunities and threats that are characteristics of the territory and environment where the household will settle down (Table 5.3). This approach leads to identifying the aspects of the project that should be reinforced or limited to adapt it to the characteristics of the territory. This distinction also enables to differentiate whether difficulties come from the household or from the territory.

Table 5.3 Three common sustainable development dimensions divided in descriptive themes and indicators

			Intern farm-focused sustainability		Extern farm-focused sustainability		Extended sustainability		Transverse themes (farm-focused sustainability)			Territorial stakes (extended sustainability)		
Sustainability dimension	General themes	Indicators	Strength	Weakness	Opportunity	Threat	Strong point	Weak point	…	Tt n°i	…	…	Et n°i	…
Socio-Territorial	A1	A1-1												
		A1-2												
	…	…												
	A12	A12-1												
		A12-2												
Economical	B1	B1-1												
		B1-2												
		B1-3												
		B1-4												
	…	…												
	B6	B6-1												
Agro-Environmental	C1	C1-1												
		C1-2												
		C1-3												
	…	…												
	C10	C10-1												
		C10-2												

Each indicator refers to one or many farm-focused or extended sustainability stake (coloured box)

For example, we can detail the A1 theme: "contribution of the household to local life". This theme is divided into two indicators A1-1 and A1-2: "involvement in associative life" and "involvement in politics". The active participation of the household members to associative or political activities represents strength for the activity system because it traduces and causes a social recognition and a territorial insertion. For the household, it is a source of motivation and a guarantee against isolation. It is a proof of the capacity of the household members to create and maintain a social network that allows us to think they will be able to mobilise an additional external work or to access information. It is strength. This household also contributes to the life of its territory and so to its sustainability, regarding this stake. However, the opportunity of taking part in the associative or political life does not depend only on the wishes of the household but also on the local dynamics or the goodwill of the other actors. Political life in some villages is so locked by natives that it is unreachable for newcomers and turns into a "threat". Thus, the study of this "contribution of the household to local life" theme with those different points of view constructs a global vision.

Finally, to introduce a dynamic lecture of the project, we propose five transverse themes which group various themes and the associated indicators. These transverse themes that give a global vision of the farm-focused sustainability of the project are *coherency between activities*, *territorial rooting*, *quality of life*, *autonomy* and *adaptability*.

5.5 Discussion

We have proposed a first structured tool to assess and support the sustainability of agricultural projects whether they are or not pluriactive. This tool constitutes only a stage and will be strengthened by the confrontation with accompanying experiments and other theoretical works. We mention below four improvement directions.

Extended sustainability stakes were selected from literature and surveys in order to introduce in the tool the most frequently quoted. This choice is based on the hypothesis that these most visible stakes represent goals of sustainability common to all actors of the territory. In order to test this hypothesis, it would be appropriate to put in debate the selected stakes among an actor's sample group. The steps of deconstruction and construction of the representations of sustainable development are the object of recent works. These works guide the methodological principles of a local co-construction of sustainable development indicators (Chia et al. 2009).

The transversal themes of restricted sustainability can also be discussed in the arena of local co-construction of indicators, in particular to improve the dynamic evaluation. So, the transverse themes "autonomy" and "adaptability" will benefit from recent works on the abilities of adaptation of the activity systems and on action in situation of uncertainty (Darnhofer et al. 2010). The flexibility and resilience are conditions of sustainability, but they do not guarantee against socio-economic marginalisation or environmental degradations. It is therefore necessary to question the dialectic between sustainability and adaptability (Ingrand et al. 2006). In this sense, it appears as a promising research to work on the identification and understanding of the factors of flexibility and resilience in order to develop an assessment tool of sustainability. Moreover, building indicators of adaptive capacity of activity systems will require to distinguish adaptation, that is, the reaction to an event, from change, that concerns modifications of the system with the aim of a better sustainability.

The representativeness and the selection of different themes to be added, kept or removed from the tool remains an open question. We did, indeed, include in the tool all the indicators identified during surveys. This can introduce an imbalance of certain themes in the overall assessment of the sustainability of the activity system. Some of the themes more modified by surveys run the risk of seeming over-represented only because they are easily identifiable during an interview. The number of indicators of a theme does not necessarily mean a greater importance in the sustainability of the activity system. Perhaps it only gives an account of larger variability in its forms of expression. In this case, how do we take it into account in assessing? More generally, is the qualitative approach sufficient to evaluate?

The tool as proposed today is the result of a bibliographic work enriched in the field by real situations. It must be tested and experimented in various supporting protocols so as to define what its integration in the support methods may be. Several modalities of use are possible. The tool could be considered (1) as a logbook: a connecting thread in the construction of the project's progression, asked and completed by the advisor on every meeting with the household or (2) as a "sheet link": an evaluation support for the farmer which he would fill out alone. It also could be considered (3) as an assessment tool to judge the progress and the weaknesses and strengths of the project at a key stage. In addition, the tool is built on an *ex post* evaluation and therefore must be tested *ex ante*: Is it usable as it is? How much time is needed to evaluate all topics? Are the data easily accessible *ex ante*?

5.6 Conclusion

This chapter presents the design process and the results of an assessment tool of sustainable activity systems of agricultural households, as a means for their support, especially during their farm creation phase. The tool is available to any organisation or advisor who wants to widen and structure their analysis of farming projects. It is conceived as an intermediate object support for dialogue and learning in the interactions between the advisor and the household. Therefore, it is not appropriate for the certification of agricultural households' projects.

Acknowledgements The authors wish to thank the following people who participated to this action research project in various ways: Aurélie Brès (ADEAR 11, France), Chloé Cadier (AIRDIE, France), Michel Dulcire (CIRAD, France) and Patrick Mundler (ISARA, France). This study was funded by the project Intersama (Insertion Territoriale des Systèmes d'Activités des Ménages Agricoles) in the framework of the regional development programme PSDR3 Languedoc-Roussillon (2008–2010).

References

Allaire, G., & Dupeuble, T. (2004). Des concepts aux indicateurs du développement durable: multidimensionnalité et responsabilisation. Accessed 21 janvier 2007.

Blanchemanche, S. (2000). *La combinaison d'activités professionnelles des ménages agricoles. L'exemple de l'Isère*. Paris: Université de Paris X.

Chia, E., Rey-Valette, H., Lazard, J., Clément, O., & Mathé, S. (2009). Evaluer la durabilité des systèmes et des territoires aquacoles: proposition méthodologique. *Cahiers agriculture, 18*(2–3), 103–118.

Curie, J., Hajjar, V., Marquie, H., Roques, M. (1990). Proposition méthodologique pour la description du système des activités. In: Les aspects psychosociaux du travail posté (I). Le travail humain, Vol. 53. PUF, Paris.

Darnhofer, I., Bellon, S., Dedieu, B., & Milestad, R. (2010). Adaptiveness to enhance the sustainability of farming systems: A review. *Agronomy for sustainable development., 30*, 545–555.

Deffontaine. J. -P. (2001). Ressources naturelles et développement durable en agriculture. In: NLD EA- (Ed.) Le développement durable, de l'utopie au concept: de nouveaux chantiers pour la recherche. Environnement. p. 288.

Gafsi, M., Legagneux, B., NGuyen, G., & Robin, P. (2006). Towards sustainable farming systems: Effectiveness and deficiency of the French procedure of sustainable agriculture. *Agricultural Systems, 90*(1–3), 226–242.

Godard, O., & Hubert, B. (2002). *Le développement durable et la recherche scientifique à l'INRA, rapport à Madame la Directrive Générale de l'INRA, rapport intermédiraire de mission*. Paris: INRA.

Ingrand, S., Magne, M. -A., Chia, E., Moulin, C., Dedieu, B., Lémery, B., Cerf, M. (2006). Non material resources mobilized by farmers and flexibility of livestock farming systems: Two concepts linked to decision making and sustainability. In: 57th Annual Meeting of the European Association for Animal Production, Antalya- Turkey, 17–20/09/2006.

Landais, E. (1998). Agriculture durable: les fondements d'un nouveau contrat social ? L*e Courrier de l'environnement* 33.

Laurent, C., & Mundler, P. (2006). L'accompagnement de la pluriactivité en question. "Territoires, acteurs et agricultures en Rhônes-Alpes" – Programme de recherches pour et sur le développement régional, Lyon.

Lefebvre, F., Bernard, N., Cessot, I. (2006). Les agriculteurs non aidés: installation et devenir. Un état des lieux national. Cnasea – Direction de la communication et des études, ADASEA, MSA, Paris.

MAAP (2007). Assises de l'agriculture. Note de problématique du groupe "Installation et pérennité des entreprises agricoles et agroalimentaires". Ministère de l'Alimentation, de l'Agriculture et de la Pêche, Paris.

Mundler, P. (2009). Viabilité et pérennité des "petites fermes". Enseignements tirés de deux projets de recherche. In: Séminaire INRA-Confédération Paysanne "Connaissances et reconaissance des petites fermes", Montpellier – France, 22–23 janvier 2009.

Mundler, P., Guermonprez, B., & Pluvinage, J. (2007). Les logiques de fonctionnement des petites exploitations agricoles. *Pour, 194*, 55–62.

Paul, M. (2004). *L'accompagnement: une posture professionnelle spécifique*. Paris: L'harmattan.
Paul, J. -L., Bory, A., Bellande, A., Garganta, E., Fabri, A. (1994). Quel système de référence pour la prise en compte de la rationalité de l'agriculteur: du système de production agricole au système d'activité. Recherches-système en agriculture et développement rural.
Pervanchon, F. (2004). L'arbre de l'exploitation agricole durable. Construire en groupe son projet d'agriculture durable. *Travaux et Innovations, 110*, 5–8.
Peschard, D., Galan, M. B., Boizard, H. (2004). Tools for evaluating the environmental impact of agricultural practices at the farm level: analysis of 5 agri-environmental methods. In: OECD expert meeting on farm management indicators for agriculture and the environment, New –Zealand, March 8–12th 2004.
RAD & CIVAM (2001). Diagnostic de durabilité du Réseau Agriculture Durable. Guide de l'utilisateur. Réseau Agriculture Durable – CIVAM, Chantepier, France.
Rattin, S. (2002). *Deux jeunes ménages sur cinq ont des ressources non agricoles*. Paris: Données sociales.
Vilain, L. (2008). *La méthode IDEA: indicateurs de durabilité des exploitations agricoles: guide d'utilisation*. Dijon: Educagri.

Chapter 6
Multilevel and Multi-user Sustainability Assessment of Farming Systems*

Steven Van Passel and Marijke Meul

Abstract A broad range of sustainability concepts, methodologies and applications already exist. They differ in level, focus, orientation, measurement, scale, presentation and intended end users. In this chapter, we illustrate that a smart combination of existing methods with different levels of application can make sustainability assessment more profound, and that it can broaden the insights of different end-user groups. An overview of sustainability assessment tools on different levels and for different end users shows the complementarities and the opportunities of using different methods. In a case study, a combination of the sustainable value approach (SVA) and MOTIFS is used to perform a sustainability evaluation of farming systems in Flanders. SVA is used to evaluate sustainability at sector level and is especially useful to support policy makers, while MOTIFS is used to support and guide farmers towards sustainability at farm level. The combined use of the two methods with complementary goals can widen the insights of both farmers and policy makers, without losing the particularities of the different approaches. We propose guidelines for multilevel and multi-user sustainability assessments.

Keywords Sustainability assessment • Multilevel • Sustainable value • MOTIFS • Farming systems • End-user groups

---

*This chapter has been published as Van Passel, S., Meul, M. (2012). Multilevel and multi-user sustainability assessment of farming systems. *Environmental Impact Assessment Review 32*, 170–180.

S. Van Passel (✉)
Centre for Environmental Sciences, Hasselt University,
Agoralaan, Building D, 3590, Diepenbeek, Belgium
e-mail: steven.vanpassel@uhasselt.be

M. Meul
Department of Biosciences and Landscape Architecture, University College Ghent,
Campus Schoonmeersen, Building C, 9000, Ghent, Belgium
e-mail: marijke.meul@hogent.be

A.A. Marta-Costa and E. Silva (eds.), *Methods and Procedures for Building Sustainable Farming Systems: Application in the European Context*,
DOI 10.1007/978-94-007-5003-6_6,

6.1 Introduction

Sustainability assessment is viewed as an important and necessary step to aid in the shift towards sustainability (Poppe et al. 2004). We need to consider which trajectories are equitable, economically and ecologically desirable and achievable (Moffatt 2000); hence, the measurement of sustainability is a daunting task. Very different sustainability evaluation tools already exist such as monetary tools, biophysical models and sustainability indicators. Examples of monetary tools are Cost Benefit Analysis (e.g. Costanza et al. 1997), the Index of Sustainable Economic Welfare (Daly and Cobb 1989) and the Genuine Savings (Pearce and Atkinson 1993). Examples of biophysical models are emergy (Odum 1996), exergy (Bastianoni et al. 2005; Hoang and Rao 2010) and the ecological footprint (Wackernagel and Rees 1997). Well-known examples of sustainability indicator sets are developed by the UN (United Nations 2001), OECD (OECD 2006) and the EU (European Commission 2005). Note that certain monetary and biophysical tools (e.g. the ecological footprint) can be identified as a kind of composite index of sustainability indicators. Furthermore, also combinations of physical indicators with monetary valuation can be identified (Neumayer 2003). An example of such a hybrid approach is the sustainability gaps approach (Ekins and Simon 1999). Interesting reviews of approaches for assessing the progress towards sustainability can be found in Neumayer (2003) and Gasparatos et al. (2008).

These different ways of measurement have been proposed regarding the monitoring and evaluation of sustainability based on different spatial, temporal and theoretical concerns (Kondyli 2010). Many sustainability assessment approaches are designed for assessments at a specific level (e.g. firm level) and are not suited to be applied at a different level (e.g. sector level) (Dantsis et al 2010). Hence, a plurality of methods is required for obtaining a sound, implementable, case- and system-specific sustainability assessment at different levels (Gasparatos et al. 2008; Hacking and Guthrie 2008; Binder et al. 2010).

The concept of scale is of major importance with regard to sustainability assessment. The term scale refers to the spatial, temporal, quantitative or analytical dimensions used by scientists to measure and analyse objects and processes (Gibson et al. 2000). Levels refer to locations along a scale, as discussed by Gibson et al. (2000). In most cases, sustainability assessment takes place at a specific level (e.g. firm level) to support decision making by a specific end-user group (e.g. firm managers). A possible shortcoming of these one-level evaluations is that the multilevel hierarchy is not considered. For example, a production unit (e.g. a firm) is always part of a production chain, so measures taken to improve the sustainability at the level of the firm will have an effect on the whole chain. A firm also belongs to an economic sector, for example, a dairy farm belongs to the dairy sector, so (policy) decisions made at sector level have an effect on the actions that can or have to be taken at firm level. Hence, performing a sustainability evaluation at the same time at different levels for different end users could broaden the insights of these different end users and provide a better support in decision making at each of the considered levels. That way, current or intended actions at, for example, the firm level most likely also contribute to the sustainability of the larger system, production chain, sector or society as a whole. In that case, instead of striving for the construction of one complete sustainability assessment approach for all levels, we propose a smart combination of existing methods applied at different levels and for different end users. Although multilevel and multi-user sustainability assessment is relevant for all kinds of systems, the literature review and case study in this chapter will be restricted to farming systems.

Many methodological approaches regarding sustainability assessment in agriculture have been published with several advantages, disadvantages and limitations (Dantsis et al. 2010). The most common approach to assess the impact of environmental or policy changes on sustainability relates to the use of indicators (Bell and Morse 1999; Diaz-Balteiro and Romero 2004; Ewert et al. 2009). The value of a sustainability indicator is its potential to improve decision making, and so it is best thought of as a source of information (Pannell and Glenn 2000). Hence, indicators describe (complex) phenomena in a quantitative way by simplifying them in such a way that communication is possible with specific target groups (Lenz et al. 2000). Furthermore, Shields et al. (2002) argue that indicators of sustainability will only be effective if they support social learning by providing users with information they need in a form they can understand and relate to. Sustainability indicators serve as performance indicators in the sense of saying to us that things are getting better or that things are getting worse (Patterson 2006). This implies that a reference point or benchmark system is necessary. To give guidance towards sustainability, reference values are needed for each indicator; these can include policy targets, best available technologies and comparisons with other countries or firms.

For agriculture, several indicator-based monitoring tools already exist and are applied in practice. These indicators generally are used (i) individually, (ii) as part of a set or (iii) combined into a composite index (Farrell and Hart 1998). Since individual indicators are of limited use to adequately represent all essential aspects of a complex system's sustainability, a balanced set of indicators is preferred (Bossel 1999). Although unconnected indicators encourage the fragmented view, combining several indicators can be seen as a significant first step to adequately assess the sustainability of an activity or firm (Farrell and Hart 1998). The next important step is to analyse the links between social, environmental and economic aspects.

Table 6.1 gives an overview of common and recent indicator systems for sustainability measurement of agricultural systems, found through a literature search in scientific journals. The literature review shows that existing indicator tools can be categorised according to the intended level of application (farm level, sector level and regional level) and the intended end-user group (farmers and policy makers). Note that it only makes sense to compare different levels if these levels belong to the same scale (Gibson et al. 2000). The intended level of application belongs to two different scales: (i) a 'production' scale (with two levels: farm level and sector level) and (ii) a 'spatial' scale (with two levels: farm level and regional level). As a consequence, analysis on sector level and spatial level cannot be compared or should be compared very cautiously (as indicated in Table 6.1 with the dotted line).

Table 6.1 Integration tools to assess sustainability at different levels for different end users

	Farmers	Policy makers
Farm level	Lewis and Bardon (1998)[a]	Andreoli and Tellarini (2000)[b]
	Girardin et al. (2000)[a]	Sands and Podmore (2000)[b]
	Ten Berge et al. (2000)[b]	Reinhard et al. (2000)[b]
	Rigby et al. (2001)[a]	De Koeijer et al. (2002)[b]
	Lopez-Ridauro et al. (2002)[a]	Pacini et al. (2004)[b]
	Hani et al. (2003)[a]	Coelli et al. (2007)[b]
	Van Calker et al. (2004, 2006)[b]	Van Passel et al. (2007, 2009)[b]
	Langeveld et al. (2007)[a]	
	Van Cauwenbergh et al. (2007)[a]	
	Meul et al. (2008)[a]	
	Rodrigues et al. (2010)[b]	
Sector level		Andreoli and Tellarini (2000)[b]
		Stoorvogel et al. (2004)[b]
		Van Passel et al. (2009)[b]
		Azad and Ancev (2010)[b]
Regional level		Smith et al. (2000)[a]
		Schultink (2000)[a]
		Stoorvogel et al. (2004)[b]
		Ewert et al. (2009)[b]
		Azad and Ancev (2010)[b]
		Balana et al. (2010)[b]
		Dantsis et al. (2010)[b]
		Hoang and Rao (2010)[b]

[a] Refers to a visual integration approach
[b] Refers to a numerical integration approach

With regard to end-user groups, we categorised the tools based on the intended or most important end users: farmers (including farm consultants) and policy makers. In certain cases, the authors claim that the analysis is useful for both farmers and policy makers (e.g. Langeveld et al. (2007)), but we tried to identify the most important target group (or end-user group). The end-user group 'researcher' is not added because we assume that all tools are also described for other researchers for further research. Certain assessment tools incorporate the perception of different stakeholders, notwithstanding the fact that these tools are used to support a certain user group. For example, Van Calker et al. (2004) take into account the perception of different stakeholders (producers, consumers, policy makers and farms) using different weights for sustainability aspects to compare dairy farming systems to support farmers as end users.

Table 6.1 shows that for policy makers as the intended end-user group, several tools exist that are used to assess sustainability at different levels. Examples are Andreoli and Tellarini (2000) who perform a farm assessment and compare different production types (or subsectors) and Van Passel et al. (2009) who perform a farm assessment and evaluate basic policy options. Stoorvogel et al. (2004) compare different production systems and analyse their spatial variation, and Azad and Ancev (2010) calculate the environmental performance index to compare production types and regions. Such comparisons cannot be considered as multilevel sustainability assessment due to the fact that different scales are considered (Gibson et al. 2000). Note that for the review in Table 6.1, we consider farm level including field level, sector level including production system level and regional level including land unit scale and (supra)national level.

An interesting and logical insight considering Table 6.1 is the fact that tools for farmers as end users are designed on farm (or field) level, while tools for policy makers as end users exist on farm, sector and regional level. Sustainability assessment comparing different regions or different production systems is logically not that useful for farmers to improve their management towards a higher sustainability performance. On the other hand, policy makers need information with regard to sustainability on different levels to support policy making on these different levels. Policy measures differ, for example, with regard to different production systems (e.g. best available techniques requirements) or different areas (e.g. restrictions in vulnerable regions).

Table 6.1 gives also an overview of both visual and numerical tools to assess sustainability, indicated with superscripts 1 and 2, respectively. To combine sustainability indicators, one can keep the indicators entirely separate, but list or present them together within a single table or diagram (visual integration), or one can combine the indicators to yield a single index of sustainability (numerical integration). In visual integration tools, sustainability indicator sets are placed in diagrammatic formats. A graphical presentation of multiple indicators allows for a comprehensive overview and mutual comparison of the indicators for different sustainability aspects. Examples of such a visual integration are radar graphs (e.g. Rigby et al. 2001; Meul et al. 2008) or bar graphs (Lewis and Bardon 1998). Graphical methods can be useful as decision aid tools, for example, to measure and compare farm progress towards a more sustainable agriculture, and they are considered well suited for effective communication about sustainability. A potential problem with visual integration tools is hidden non-linearities or interactions between indicators. Most of the indicators within a set have not been linked together, although sometimes trade-offs among issues exist that cannot be resolved simultaneously (Cornelissen et al. 2001). Furthermore, from some indicator lists, de Haan (2004) gets the impression of a fairly incoherent shopping list of numbers without underlying structure. Also, Farrell and Hart (1998) argue that in many cases, the sustainability indicators are simply combined lists of traditional economic, environmental and social indicators with the word sustainable added to the title. A final potential problem of indicator sets is that often a large number of indicators giving information on developments in the economic, social and environmental areas and including both qualitative and quantitative factors are used, resulting in lists that are often long and impractical in use (Lopez-Ridauro et al. 2002). However, recent visual integration tools are in general user-friendly and a communicative instrument to measure progress towards sustainability.

A different way to aggregate sustainability indicators is the numerical approach where a composite indicator is constructed by combining different components into one single unit. Composite indicators can be defined as based on sub-indicators that have no common meaningful unit of measurement, and there is no obvious way of weighting these sub-indicators. Different methods can be used to compose such an indicator such as efficiency analysis (e.g. Reinhard et al. 2000; De Koeijer et al. 2002; Coelli et al. 2007; Azad and Ancev 2010), the sustainable value approach (e.g. Van Passel et al. 2007, 2009), modelling approaches (e.g. ten Berge et al. 2000; van Calker et al. 2004; Pacini et al. 2004; Stoorvogel et al. 2004) and multi-criteria analysis (e.g. Andreoli and Tellarini 2000; Balana et al. 2010; Dantsis et al. 2010). Aggregated sustainability indicators in a compact form are in particular useful to compare policy options (Farrell and Hart 1998), because they summarise complex or multidimensional issues and they provide the big picture (Saisana et al. 2005), without the danger of information overload. Furthermore, aggregated indices can help to convey simple messages and to reach new audiences but also run the risk of being misinterpreted. The lack of transparency by highly aggregated indicators can be a serious problem (Bell and Morse 2003). Therefore, it is essential that these indices satisfy several quality criteria and are interpreted in their proper context. Jollands et al. (2004) conclude that aggregate indices do have a role in assisting decision makers, as long as they are not used in isolation from more detailed information. Costanza (2000) notes that detailed information of aggregated indicators is not lost; usually, it is possible to look at the details of how any aggregate indicator has been constructed, but decision makers are too busy to deal with these details. Sauvenier et al. (2005) argue that the aggregation of indicators is a net advantage; since indicators are a prerequisite to aggregation, the most detailed information always stays available. Kondyli (2010) even states that the creation of robust composite indicators is an imperative due to their comprehensiveness and ease of communication and interpretation as an appealing tool for policy makers. An interesting overview of the pros and cons of composite indicators for the evaluation of agricultural sustainability is presented by Gomez-Limon and Sanchez-Fernandez (2010).

Table 6.1 clearly shows that visual integration tools are mostly used on farm level to support farmers as end users, while numerical integration tools are mostly used to support policy makers as end users. Visual integration tools assessing sustainability on multiple levels remain scarce, while numerical integration tools are sometimes used on different levels (e.g. Andreoli and Tellarini 2000; Stoorvogel et al. 2004; Van Passel et al. 2009; Azad and Ancev 2010). A related and interesting aspect is the fact that several tools use farm-level data to assess sustainability on sector or regional level. Examples are Andreoli and Tellarini (2000), Ewert et al. (2009), Van Passel et al. (2009) and Dantsis et al. (2010). In fact, farm-level data are aggregated in a certain way to assess sustainability on a different level.

In summary, the literature review shows that the two considered end-user groups (farmers and policy makers) set different requirements for a sustainability evaluation tool. Visual integration tools using farm data are most appropriate to inform farmers on the sustainability at farm level. Policy makers on the other hand have more benefit from using numerical integration tools applied at farm, sector or regional level, where farm data are often used to assess sustainability at different levels. These findings lead us to the consideration that a multilevel and multi-user sustainability assessment can benefit from combining different tools with specific designs instead of trying to develop one complete sustainability assessment approach for all end users at all levels.

To expand this general concept of combining sustainability assessment tools, we perform a sustainability evaluation of farming systems simultaneously at farm level and sector level to inform both farmers and policy makers by combining two existing sustainability assessment tools: sustainable value approach (SVA) and MOTIFS (monitoring tool for integrated farm sustainability). Both methods provide good guidance for decision making and make sustainability operational in a clear way, but for different end users and at different levels with regard to the same scale ('production' scale). SVA is used to allow policy makers to compare sustainability performance of different agricultural sectors, while MOTIFS allows farmers to measure the progress towards sustainability at farm level.

In a case study, SVA is used to compare the sustainability of the Flemish specialised arable and dairy sector. At the same time, at farm level, MOTIFS is used to guide individual farmers within their subsector (e.g. dairy farming) in taking the proper actions towards more sustainable farms.

In the following section, we present the case study where SVA and MOTIFS are used simultaneously to perform a sustainability evaluation of agricultural systems in Flanders, using data of 28 specialised farms. In Sect. 6.3, we propose guidelines for multilevel and multi-user sustainability assessment based on the experiences of the case study and the categorisation of the existing sustainability monitoring tools for agricultural applications according to their intended level of application and intended end users in this section (see Table 6.1). A final section concludes.

6.2 Case Study: A Practical Multilevel and Multi-user Sustainability Assessment Using SVA and MOTIFS

6.2.1 Sustainable Value Approach (SVA)

SVA is developed by Figge and Hahn (2004, 2005) and applies the logic of opportunity costs to the valuation of resources using a capital approach (e.g. Atkinson 2000). Using SVA, we consider that a firm contributes to more sustainable development whenever it uses its resources (economic, environmental and social) more productively than other companies and the overall resource use is reduced or unchanged. The following steps are required to calculate the sustainable value of a company:

- First, the scope of the analysis needs to be determined (i.e. economic activity/activities or entity/entities).
- Second, the relevant critical corporate resources with regard to sustainability performance within the chosen scope need to be determined.
- Third, the benchmark value needs to be determined. The choice of the benchmark determines the cost of the resource needs of a company, in other words the

productivity that a company has to exceed. An interesting methodological and conceptual discussion about using benchmarks to measure the sustainable value can be found in Kuosmanen and Kuosmanen (2009), Figge and Hahn (2009) and Ang and Van Passel (2010).

- Then the productivity (also referred to as eco-efficiency when related to environmental resources) of a certain corporate resource is compared to the one of the benchmark while keeping the overall resource use constant. If the productivity of the firm exceeds the opportunity cost (productivity of the benchmark), the company contributes to sustainability for the resource concerned.
- The differences between the company and the benchmark productivity are then summed up for all relevant resources and divided by the amount of considered resources with the sustainable value (SV) as a result (Figge and Hahn 2004, 2005). To take the firm size into account, a return-to-cost (RtC) ratio can be calculated by dividing the value added by the cost of the sustainable capital. The cost of sustainable capital is given by the difference between the value added and the sustainable value (Figge and Hahn 2004; Van Passel et al. 2009).

SVA is a useful tool to formulate advice and assist policy makers in decision making at different levels. The sustainable value summarises the sustainability performance of a company, activity or sector into one single value, while the productivity results of the individual resources are still easily available for a more detailed policy interpretation of the results. The sustainable value approach is already used for several interesting applications such as the sustainability assessment of an oil company (Figge and Hahn 2005), the European manufacturing companies (Hahn et al. 2007), the automobile industry (Hahn et al. 2009), German companies (Hahn et al. 2010), farms (Van Passel et al. 2007, 2009) and European countries (Ang et al. 2011).

Note that the sustainable value approach can be used on both farm level and sector level. Van Passel et al. (2009) show how to use farm-level data to analyse the evolution and the impact of policy decisions on sector level (using average values). However, we stress that in all applications (firm and sector level), the most important end users of the sustainable value approach are the policy makers.

6.2.2 Monitoring Tool for Integrated Farm Sustainability (MOTIFS)

MOTIFS is an indicator-based sustainability tool to monitor farm progress towards integrated sustainability, that is, taking into account economic, ecological as well as social aspects. The tool offers a visual integration of indicator scores into an adapted radar graph, considering ten sustainability themes related to ecological, economic and social aspects (Fig. 6.1). To aggregate the indicators for different sustainability themes, benchmarks were defined to rescale indicator values into scores between 0 (indicating a worst-case situation) and 100 (indicating assumed maximum sustainability). This allows for a comprehensive overview and mutual comparison of the indicators for different sustainability themes. MOTIFS is a visual monitoring tool. Starting from an overall view of his farm's sustainability, a farmer can zoom in on

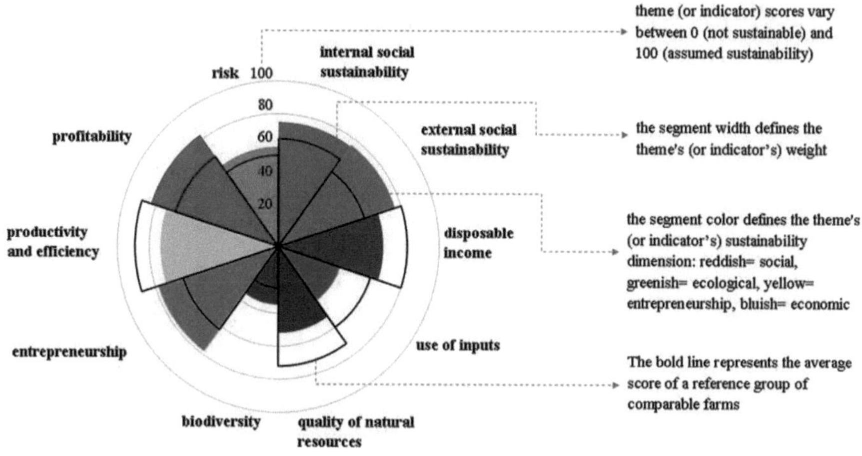

Fig. 6.1 Monitoring tool for integrated farm sustainability (MOTIFS), presented with a legend concerning the reading and interpretation

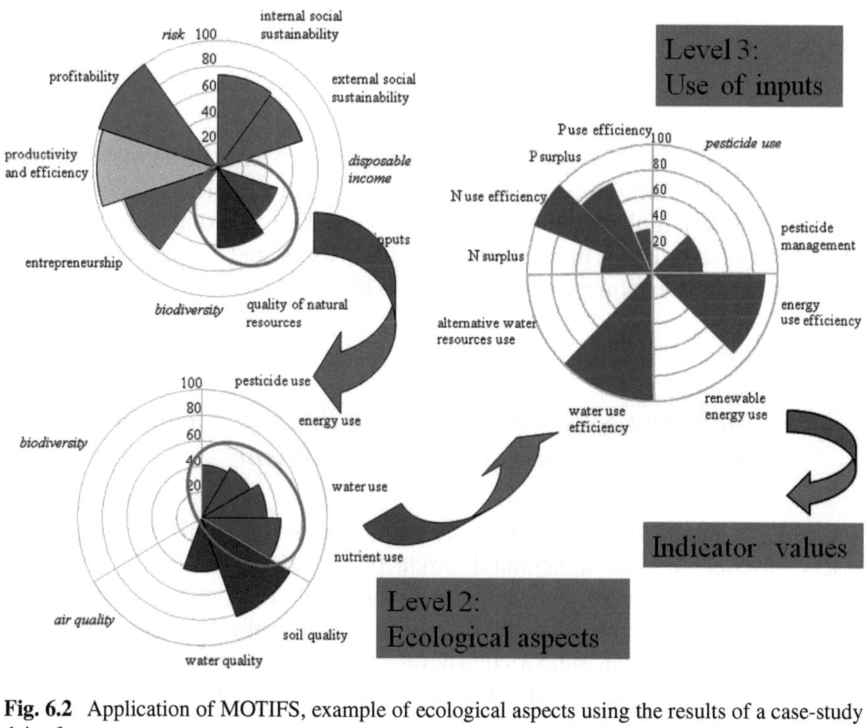

Fig. 6.2 Application of MOTIFS, example of ecological aspects using the results of a case-study dairy farm

the underlying themes and indicators into as much detail as desired. This is shown in Fig. 6.2. A detailed description of MOTIFS and its underlying methodology are provided by Meul et al. (2008).

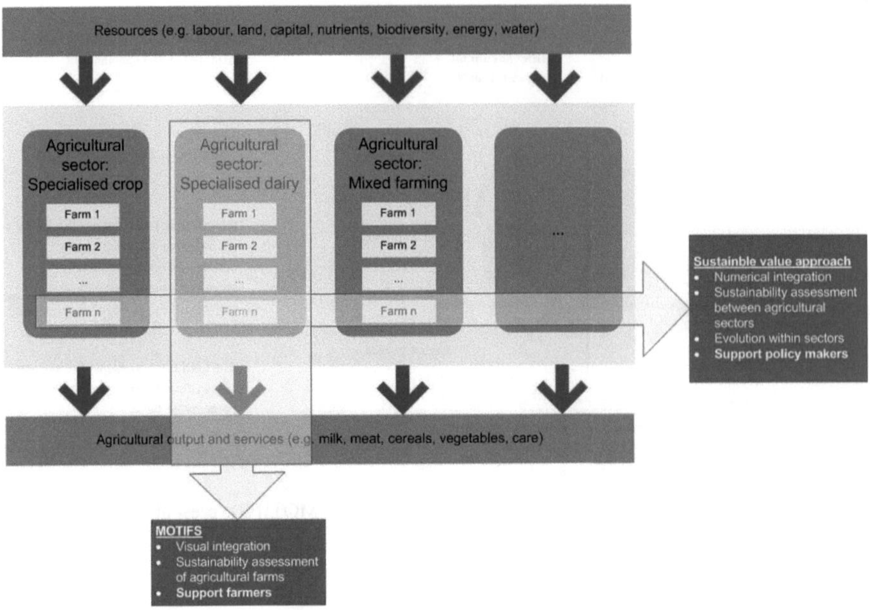

Fig. 6.3 Framework for a multilevel and multi-user sustainability assessment of farming systems

The aim of MOTIFS is to guide farmers' management towards higher sustainability. The optimisation of MOTIFS as a social learning tool and as a sustainability management tool has been further examined by De Mey et al. 2011 and Marchand et al. 2010. MOTIFS is already used in different projects with advisors' and farmers' organisations to evaluate and guide farm management towards higher sustainability.

6.2.3 Combining SVA and MOTIFS to Perform a Multilevel and Multi-user Sustainability Assessment

Figure 6.3 shows an illustration of a practical multilevel and multi-user approach to assess sustainability of agricultural production systems combining SVA and MOTIFS. The sustainable value of different agricultural activities is calculated to evaluate and compare the performance of different agricultural subsectors. In this way, policy makers can be supported to develop a well-balanced and focused policy. Simultaneously, the sustainability of farms within a specific subsector can be monitored using MOTIFS. In other words, SVA is used to assess sustainability on sector level (comparing different agricultural subsectors), and MOTIFS is used to assess sustainability on farm level (within a particular agricultural subsector). On the other hand, the SVA results are useful to support policy makers, and the MOTIFS analysis is useful to support farmers. Hence, the combination of MOTIFS and SVA results in a practical multilevel and multi-user sustainability assessment.

Combining SVA and MOTIFS is feasible because they are both based on a similar sustainability concept. MOTIFS is founded on the equality of the economic, ecological and social sustainability dimension, while SVA integrates environmental, social and economic aspects into a monetary analysis based on opportunity costs. So, in fact, SVA is also based on the three-pillar approach inventorying environmental, economic and social resources. Note, however, that the three-pillar approach is inherently built into MOTIFS, while in most past SVA studies, a considerable weighting towards environmental resources can be observed (Ang and Van Passel 2010). On the other hand, both methods approach sustainability from a different point of view. SVA envisions sustainability from a resource use perspective, while MOTIFS translates major principles of a supported vision on sustainable agriculture (Nevens et al. 2008) into concrete and relevant themes. The applied methodology of MOTIFS hence fits within a content-based framework (von Wirén-Lehr 2001). This means that specific sustainability aspects, for example, animal welfare, are considered in MOTIFS but are not seen as a resource in SVA. When combining MOTIFS and SVA, we should make sure that the resources that are considered in SVA are also considered in MOTIFS to avoid that the sustainability evaluation at both levels (farm and sector level) is based on different sustainability aspects. In our case study, all resources considered in SVA are also evaluated with MOTIFS. This results in an evaluation that is mainly economic and ecological. For practical reasons, the social sustainability aspects are not considered in the case study since social resources could not easily be retrieved from the farm accountancy data.

Note that the motivation to opt for a combination of SVA and MOTIFS is rather pragmatic. Also, other tools are complementary with regard to level and end user (see Introduction). However, both SVA and MOTIFS are already used and validated on Flemish farms (see Van Passel et al. 2007, 2009 and Meul et al. 2008, 2009).

6.2.4 Case-Study Farms

Farm accountancy data from specialised dairy and arable farms in Flanders (Belgium) are used for both SVA and MOTIFS. These data were collected by the European FADN database (Farm Accountancy Data Network). The Flemish FADN data are collected and managed by the monitoring division of the Agricultural Monitoring and Study service of the Flemish Ministry for Agriculture. Both technical and economic data from a representative set of Flemish farms are available. We considered dairy farms as 'specialised' when at least 95 % of the farm income originated from dairy activity. Specialised arable farms get at least 95 % of the farm income from arable production. An overview of some average descriptive characteristics of the selected farms is presented in Tables 6.2 and 6.3.

Table 6.2 Average descriptive statistics of the data sample of specialised Flemish arable farms (Data of the year 2000)

Characteristic	Unit	Average value
Arable farms	#	14
Cultivated area	ha	34.4
Share of grain crops[a]	%	46.0
Share of potatoes[a]	%	16.1
Share of sugar beet[a]	%	24.9
Share of maize[a]	%	10.3
Yield of winter wheat	kg ha^{-1}	7,488
Yield of sugar beet	kg ha^{-1}	64 486
Yield of potatoes	kg ha^{-1}	37 235
Value added	€	24 091
Labour	h year^{-1}	2,825
Farm capital	€	161 467
Energy use (direct and indirect)	MJ	755 147
N surplus	kg N ha^{-1}	139
Age manager	year	50
Higher education	%	36

[a]As a percentage of the total cultivated area

Table 6.3 Average descriptive statistics of the data sample of specialised Flemish dairy farms (Data of the year 2000)

Characteristic	Unit	Average value
Dairy farms	#	14
Utilised area	ha	34.1
Milking cows	#	56
Milk production	l cow^{-1} year^{-1}	6,350
	l ha^{-1} year^{-1}	11,380
Value added	€	54,120
Labour	h year^{-1}	4,450
Farm capital	€	5,73,341
Energy use (direct and indirect)	MJ	12,89,397
N surplus	kg N ha^{-1}	288
Age manager	year	41
Higher education	%	80

6.2.5 Sustainable Value Approach: Dairy Farms Versus Arable Farms

Considering the different steps of SVA as described in Sect. 6.2.1, the following choices were made:

- The scope of the analysis can be described as the Flemish dairy and arable specialised farming sector.

- As in Van Passel et al. (2007, 2009), we consider five resources in the sustainability evaluation: (i) farm labour, (ii) farm capital, (iii) farm land, (iv) nitrogen surplus and (v) energy consumption (direct and indirect). Capital, land and labour can be seen as traditional economic resources, while nitrogen surplus and energy consumption are important environmental aspects for Flemish farms (Nevens et al. 2006; Meul et al. 2007). Note that for both specialised dairy farms and arable farms, the same relevant resources are selected. These resources were chosen based on the availability of data within the FADN dataset. The data sample contains dairy farms and arable farms with similar regional, soil and land use characteristics. In this way, a general comparison with regard to the sustainability performance (based on the selected resources) is possible. The value added is used as outcome measure.
- In our application, we opt as benchmark for the average return on resource use of the whole sample of 28 farms (14 dairy and 14 arable farms). We use the original benchmark technology, because our aim is to present the overall resource efficiency of the farm from the investor's viewpoint (Ang and Van Passel 2010). Moreover, implementing productive efficiency theory to benchmark would require more observations.
- For each farm and each resource, the firm productivity (or eco-efficiency) is compared with the productivity of the proposed benchmark.
- The differences between the farm and the benchmark productivity are then summed up for all relevant resources and divided by five (i.e. the amount of considered resources) to calculate the sustainable value (SV) for each farm. In fact, the sustainable value estimations indicate how much more or less return each farm creates with the resources available in comparison with the benchmark. Furthermore, the RtC (return-to-cost) ratio is calculated to take the differences in farm size into account. A RtC higher than one means that the company is overall more productive than its benchmark. The return-to-cost ratio shows by which factor the farm exceeds or falls short of covering its cost of economic, environmental and social resources or in other words by which factor it exceeds or falls short compared with the benchmark productivity. The average RtC of both the specialised dairy and arable farms is used to compare the sustainability performance on sector level.

Figure 6.4 shows the return-to-cost ratio of the specialised dairy (14) and arable farms (14). The performance of the arable farms is on average lower than the performance of the dairy farms (RtC: 0.89 versus 1.05). However, the two best-performing farms are arable farms. Figure 6.4 also shows that larger differences exist within the arable farms (range RtC: 0.1–2.1) compared to the dairy farms (range RtC: 0.4–1.6). Note that this analysis is rather descriptive, and one should be careful with generalisation of the results.

The differences in average resource productivities and eco-efficiencies between the arable and dairy farms can be found in Table 6.4. We see that dairy farms have a very high land productivity compared to arable farms and a higher labour productivity and eco-efficiency of the energy use. On the other hand, arable farms clearly outperform dairy farms with regard to capital productivity and eco-efficiency of N

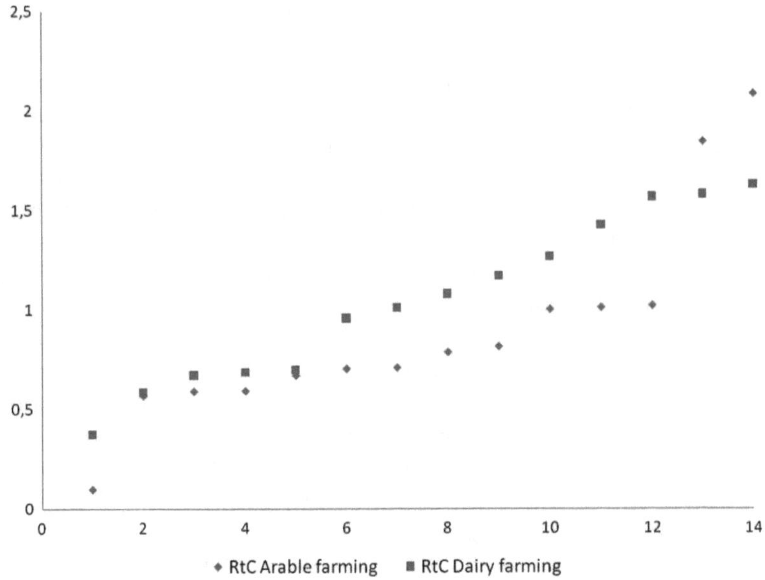

Fig. 6.4 Return-to-cost ratio using average benchmarks

Table 6.4 Average resource productivities and eco-efficiencies

	Labour productivity (€/hour labour)	Capital productivity (€/€)	Land productivity (€/ha)	Eco-efficiency energy use (€/MJ)	Eco-efficiency N surplus (€/kg N)
Arable farms	9.17	0.18	713.48	0.03	9.37
Dairy farms	11.3	0.10	1568.94	0.04	6.21

A one-way ANOVA test shows that the average capital and land productiveness differ significantly between arable and dairy farms (F-value > 4.23)

surplus. From these SV calculations, it can be advised that a clear focus on the reduction of the N surplus on dairy farms is important to strengthen the sustainability performance of the Flemish dairy sector. Arable farming in Flanders has clear limitations due to space constraints. However, increasing labour productivity and value added are possible as shown by two arable farms with a high RtC (Fig. 6.4).

In other words, a straightforward policy conclusion comparing the agricultural subsectors is to focus on the reduction of N surplus of dairy farms (e.g. feed optimisation) and to focus on higher value creation on arable farming (e.g. on farm sales).

Table 6.5 shows the differences of farm characteristics between high- (RtC > 1) and low- (RtC ≤ 1) performing farms. The majority of considered arable farms have a RtC ≤ 1, while the majority of the dairy farms have a RtC > 1. For the resources considered, sustainability performance of Flemish specialised dairy farms is in general higher compared to the performance of specialised arable farms. In general, younger farm managers obtain better results, while education and solvency have no impact

Table 6.5 Average descriptive statistics of all farms, frontrunners and laggards

Variables	All farms	Farms with a RtC ≤ 1	Farms with a RtC > 1
Sustainable value (Euro)	0	-14,156	+14,157
Age of manager (years)	45	49	42
Solvency[a] (%)	65	64	67
Share higher education (%)	57	57	57
Share arable farms (%)	50	64	36

[a]Measured as own capital divided by total capital

on the sustainability performance. A straightforward policy suggestion would be to stimulate succession and to rejuvenate farming towards specialised dairy activities. Note that to explain the differences in RtC in more detail, an econometric panel data should be estimated (as in Van Passel et al. (2007)). With regard to this application, not enough observations were available.

6.2.6 MOTIFS: Sustainability of Dairy Farms and Arable Farms

In the application of MOTIFS, individual farm performances are compared within each agricultural subsector, contrary to the application of the SVA method, where the dairy farms and arable farms are compared to one another. As an example, we describe the results of the MOTIFS application to the selected dairy farms; the application and use of the MOTIFS to arable farms are completely similar. For the 14 specialised dairy farms, the following MOTIFS indicators were calculated based on the FADN data: (i) N surplus, N use efficiency and direct and indirect energy use efficiency, to evaluate the ecological sustainability of the farms and (ii) productivity and profitability indicators to evaluate their economic sustainability. An overview and description of these indicators can be found in Nevens et al. (2006), Meul et al. (2007) and Meul et al. (2008).

Table 6.6 shows the results of the sustainability evaluation of the 14 dairy farms based on the selected indicators. For each indicator, the lowest, highest and average values are shown. These indicator values were converted into a score between 0 and 100 for each indicator by using respectively the results of the lowest-performing and best-performing case-study farm as benchmark values. This choice of benchmarks was made based on the validation results, where users of the tool expressed their appreciation of using indicator values of the 10 % best-performing and 10 % lowest-performing farms as benchmarks, since this results in a dynamic and motivating tool for farmers, setting realistic goals (Meul et al. 2009). For each farm, indicator scores were integrated in a MOTIFS graph. Figure 6.5 shows the MOTIFS results for four case-study dairy farms as an example. For each indicator, the average score of a large representative group of FADN dairy farms is indicated by the black bold line.

Table 6.6 Indicator values dairy farms

		Lowest value	Average value	Highest value
Economic analysis				
Labour productivity	€/MWU[a]	7248.61	27,232.02	53,724.80
Capital productivity	€/€	0.04	0.10	0.18
Land productivity	€/ha	588.50	1568.94	2509.92
Labour profitability	€/MWU[a]	-34951.96	-1512.83	13708.18
Return on equity	€/€	-0.52	-0.17	-0.01
Return on assets	€/€	-0.15	-0.06	0.00
Ecological analysis				
N surplus	kg N/ha	144.43	287.65	598.35
N use efficiency	l milk/kg N surplus	20.62	40.90	58.28
Direct energy use efficiency	l milk/ 100 MJ direct energy use	48.73	86.48	177.61
Indirect energy use efficiency	l milk/ 100 MJ indirect energy use	30.51	41.19	58.91

[a]MWU = Man-work unit. 1 MWU is the equivalent of 2,400 working hours

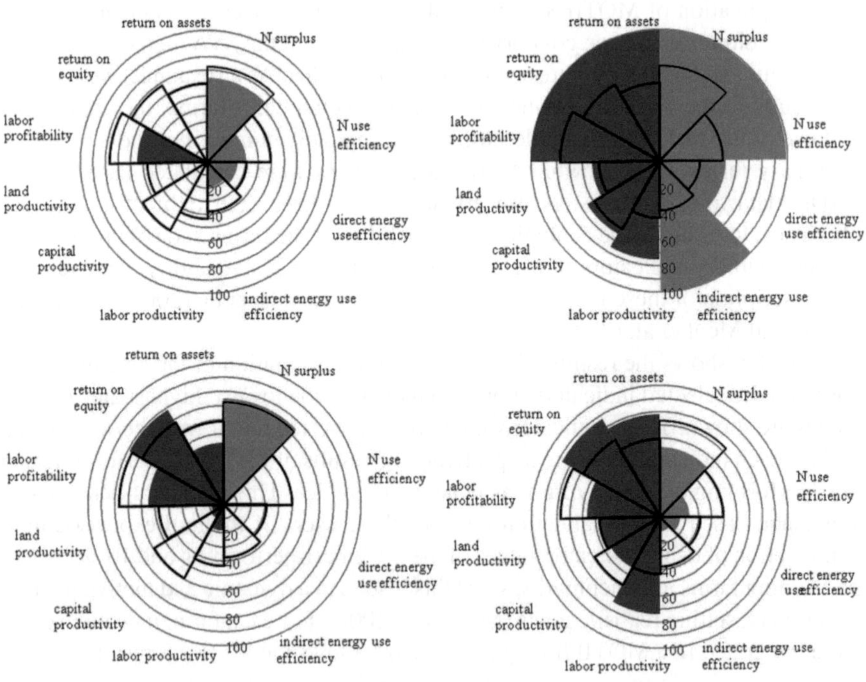

Fig. 6.5 MOTIFS results of four dairy farms

Application of MOTIFS involves the discussion of the MOTIFS results in a discussion group, in which the 14 dairy farmers would participate. During these discussions, farmers exchange knowledge and expertise and discuss the background of the indicator results with an invited expert. For example, using the results shown in Fig. 6.5, the top left farm could be set as an example for the considered economic and ecological sustainability aspects. Experiences and management practices of that farmer – combined with an expert opinion – can be an inspiration for the other farmers and give them insights into management aspects or innovations that could be applied to their own farm in order to improve the sustainability.

Recent applications of MOTIFS in similar discussion groups of farmers showed that to further improve the effective use of MOTIFS by farmers, specific attention should be given to the organisation of the discussion sessions (De Mey et al. 2011). Since discussion sessions of farmers are mostly guided by farm advisors, they should be well trained to translate indicator results into advice. Enhancing advisors' communication skills is also crucial to facilitate interactive and flexible dialogues that lead to better learning among farmers. Also, a thorough planning of the discussion sessions is necessary, with clear goals for each session. Involving experts on particular themes of MOTIFS can make the dialogues more profound and produce more tangible advice. Finally, discussion sessions of farmer groups could be combined with individual discussions between the farmer and advisor. Individual action plans can then be developed per farm and discussed in group.

6.3 Multilevel and Multi-user Sustainability Assessment of Farming Systems

6.3.1 Lessons Learned from the Case Study

In the SVA approach, the sustainable value of 14 dairy farms is compared with the sustainable value of 14 arable farms. The dairy farms in our sample have on average a higher sustainability performance than the arable farms. In other words, dairy farms realise relatively more sustainable value using their resources (both economic and ecological). Note that the sustainable value integrates the performance (productivity and eco-efficiencies) of different resources. In this case, the relatively good performance of dairy farms of using labour, land and energy outweighs the low performance of using capital and N surplus compared to arable farms. Hence, a straightforward policy advice using SVA is to stimulate the development of well-balanced dairy farms with a clear focus on reducing the N surplus.

On farm level, MOTIFS can be used to compare the sustainability performance within a discussion group of comparable farms (e.g. specialised dairy farms). In our illustration, indicator scores of the dairy farms are calculated and integrated in a MOTIFS graph. Farmers can then discuss the background of their indicator results with other farmers and experts. Farm experiences and management practices together with expert opinions motivate and stimulate farmers to improve their sustainability.

More specifically, dairy farmers can exchange practices and knowledge to reduce the amount of N surplus. On the other hand, analysing the sustainable value, dairy farmers can recognise that the N surplus of specialised arable farms is clearly lower than the N surplus of dairy farms. In the other way around, farm-level sustainability analysis using MOTIFS could help policy makers by formulating within discussion groups useful policy ideas or targets to reduce the N surplus.

SVA can be seen as a clear reductionist approach reducing a complex system into a simple number. MOTIFS uses indicators to visualise the progress towards sustainability of complex systems in an understandable way. On the other hand, MOTIFS advocates stakeholder involvement and participatory sustainability assessment. The use of both approaches (SVA and MOTIFS) can improve the mutual understanding of the different target groups, respectively policy makers and farmers. In fact, using complementary strategic approaches can generate transformational knowledge and can have the potential to promote knowledge brokerage (Sheate and Partidario 2010). Moreover, issue importance (with regard to sustainability) is a necessary but not a sufficient condition for policy attention (Engstrom et al. 2008). The presence of strong and well-organised stakeholders and limited attention of policy makers might be decisive for issue attention (Engstrom et al. 2008). Tools as MOTIFS can stimulate the formation of stakeholder networks, while tools as SVA can support policy makers.

An interesting insight is that for both methods – SVA and MOTIFS – similar data can be used. In other words, the same farm accountancy data can be used to feed different indicator systems with different complementary goals on different levels (e.g. farm level and sector level) and for different end users (farmers and policy makers). In this way, using a combined approach does not require a significantly higher amount of time and data.

6.3.2 Guidelines for Multilevel and Multi-user Sustainability Assessment

Our case study clearly shows the added value of combining sustainability assessment methods on farm and sector level. It is advisable to consider different levels when a sustainability assessment is performed. Combining methods on different levels (and developing a multilevel approach) can maintain the particularities and strengths of each method. But how should (existing) sustainability tools at different levels be combined? Can we develop a framework to be able to choose the right combination of methods for any case?

According to Gasparatos et al. (2008), incorporating sustainability assessment tools into integrated assessment frameworks can result in better guiding of sustainability planning and decision making. Conscious evaluation tool selection will reduce the risk of providing distorted sustainability evaluations (Gasparatos 2010). In addition to this, the combined use of existing methods should minimise data and time requirement to realise the higher impact expected and thus optimise the equilibrium of accuracy and practicality.

However, although several papers compare the differences of sustainability assessment tools (e.g. Hanley et al. 1999; Gasparatos et al. 2008; Binder et al. 2010), there exist few applications where approaches were combined in a complementary way. Ewert et al. (2009) state that the scientific basis for linking models across disciplines and levels is still weak and requires specific attention. Gasparatos et al. (2008) propose to consider the use of a variety of tools and to integrate the outputs from such a pluralistic approach. Binder et al. (2010) propose to combine fast, easily measurable indicators with indicators providing a more site-specific complex system perspective including stakeholders.

Moreover, Gasparatos (2010) stresses the importance to consider the human values when choosing an evaluation tool and to understand stakeholder attitudes. In fact, this consideration is even more important if different evaluation tools are combined. Mixing up of the key assumptions and objectives of the different tools should be avoided, and extra attention should be going to communication and explanation. Nykvist and Nilsson (2009) conclude that to enhance the potential for integrating sustainability concerns, it seems less fruitful to develop more advanced and complex assessment frameworks and models than strengthening institutional arenas for social learning.

Taking into account these considerations, we believe that a general framework for multilevel sustainability assessment is not feasible and desirable at this point. Such a framework can hamper the necessary diversity of case-specific sustainability assessments. Nevertheless, to stimulate and support further research and applications, we propose the following guidelines for multilevel and multi-user sustainability assessments based on our experiences and literature review:

1. Distinguish and describe the different levels and/or different end users involved. An important finding from our study is the fact that different end users of sustainability assessment tools prefer different approaches (visual versus numerical integration approach). Visual integration is often preferred to support farmers as end users, while numerical integration assessment tools are preferred by policy makers. Langeveld et al. (2007) evaluated farm performance using agri-environmental indicators and showed that indicators measuring a specific aspect (such as nitrogen or carbon) can and should differ depending on the target group (farmers, policy makers and researchers). This was also concluded by Rodrigues et al. (2010) who developed a system for integrated farm sustainability assessment with both a visual and numerical integration approach. Our literature overview also showed that sustainability evaluations at farm level are most often performed using visual integration tools, while evaluations at sector or regional level are mostly performed using numerical integration tools. In our case study, we applied a visual integration tool at farm level to inform farmers and a numerical integration tool at sector level to inform policy makers.
2. Identify the relevant sustainability aspects in a broad perspective. Many existing indicator-based monitoring tools used in agriculture focus on an economic-ecological evaluation of sustainability. Some tools also include social aspects and are hence built on the three-pillar concept of sustainability. Depending on the

goals and interests of the study, a selection of sustainability themes has to be made. In our case study, both SVA and MOTIFS are based on the three-pillar concept of sustainability; however, due to practical aspects and data availability, only the indicators related to economic and ecological sustainability (easily available from farm accountancy data) were used. Therefore, in our case study, an economic-ecological evaluation of sustainability was performed, using a selection of indicators from existing tools.

3. Select the appropriate sustainability assessment tool for each level considered, taking into account the intended end users, the considered sustainability aspects, the available access to the sustainability tool's methodology and the necessary data requirements. The tool's methodology and calculation method of the selected indicators should be well documented to allow for a clear insight into the evaluation and a solid application of the tool. Moreover, practical applicability of the selected tools is essential, and availability of reliable data at each level is important. In our case study, a clear description of the selected tools and their indicators were available through several scientific publications. Both tools used the publicly available FADN database as a data source. However, data concerning social aspects were not available, so these could not be considered.
4. Check the consistency and complementarity of the selected tools. On the one hand, the selected tools should be complementary with regard to the level of application, the intended end users and the insights the results can provide. On the other hand, the tools should be consistent with regard to the sustainability concept they are based on, their objectives and assumptions made. These assumptions relate to, for example, the conversion factors that are used to calculate the indicators, the system boundaries that are considered or the accuracy of the data at each level. In our case study, the selected tools SVA and MOTIFS were complementary concerning (i) the level of application (farm versus sector), (ii) the intended end users (farmers versus policy makers) and (iii) the insights provided (how to improve the sustainability of a sector versus how to improve the sustainability of an individual farm). The tools were consistent since they both considered the same sustainability aspects, the same system boundaries (evaluation at both levels was based on farm data considering the same boundaries), and similar data were used to perform the calculations of both tools.
5. Organise discussion sessions with the different end users. As advocated by several authors (Nykvist and Nilsson 2009; Meul et al. 2009; De Mey et al. 2011; Marchand et al. 2010), establishing arenas for social learning – such as discussion sessions with end users – provides a significant added value to the sustainability assessment and action-taking towards higher sustainability. When different end-user groups are involved, we can think that such discussion sessions involving these different end users becomes even more important as a tool for mutual understanding, guiding and learning. Concerning our case study, some practical applications of MOTIFS involving discussion sessions with farmers exist. One application, described by De Mey et al. (2011), involves both policy makers and farmers, but here, the policy makers merely have a facilitating and guiding function for the farmers. In these discussion sessions, mutual learning only takes place

between farmers (at farm level), guided by an expert and organised by the policy makers. De Mey et al. (2011) investigated the social learning between farmers based on the MOTIFS results and identified some critical success factors. An application of SVA at sector level and MOTIFS at farm level followed by a discussion session including farmers and policy makers did, however, not yet take place.

6.4 Conclusion

In this chapter, we explored the possibilities to perform a multilevel and multi-user sustainability evaluation of farming systems through a smart combination of existing sustainability monitoring tools with different purposes and applied at different levels and for different end users. Performing a sustainability evaluation at different levels at the same time could broaden the insights and provide a better support in decision making at each of the considered levels so that current or intended actions at, for example, the firm level most likely also contribute to the sustainability of the larger system, production chain, sector or society as a whole (and vice versa).

To achieve this goal, we performed a literature review of existing sustainability assessment tools in agriculture and classified them according to their level of application and intended end users. An interesting finding from this literature review was that visual integration tools are mostly used on farm level to support farmers as end users, while numerical integration tools are mostly used to support policy makers as end users.

In a second step, we performed a case study where two sustainability monitoring tools were used simultaneously to perform a sustainability evaluation of agricultural systems in Flanders at farm level and sector level: SVA (a numerical integration of sustainability dimensions) was used to assess sustainability on sector level supporting policy makers, while MOTIFS (a visual integration tool) was used to guide farmers towards sustainability on farm level. Moreover, SVA can be seen as an easy and fast policy assessment, while MOTIFS provides a more detailed and specific system perspective including stakeholders. The application showed that using SVA combined with MOTIFS results in a sustainability multilevel assessment, strengthening and framing the assessment on the different levels without losing the particularities of the different approaches. The combination of methods with different aims made sustainability assessment more profound and broadened the possible insights. The application showed that a combination of numerical and visual integrated sustainability measurements can lead up to clearer insights and to a more effective use of the sustainability assessment tools. In other words, a broader and more profound sustainability measurement can motivate and recruit more users (decision makers).

It was our aim to develop a general framework for combining existing monitoring tools to perform a multilevel and multi-user sustainability assessment. However, considering the diversity of case-specific approaches for sustainability assessment, we did not find it possible nor opportune to describe a general framework for combining the existing methods. Nevertheless, to stimulate and support further research and applications, we proposed guidelines for multilevel and multi-user sustainability assessments based on our experiences and literature review. A multilevel sustainability assessment requires a clear identification of the needs of the different levels and end users involved and the relevant sustainability aspects in a broad perspective. The most appropriate integration approach (visual or numerical integration) for each level and type of end user should be selected. The consistency and complementarity of the different assessment tools should be verified. Furthermore, the collection and processing of data, feeding the different sustainability assessment tools, should be optimised. Finally, discussion groups involving the different end-user groups should be organised.

References

Andreoli, M., & Tellarini, V. (2000). Farm sustainability evaluation: Methodology and practice. *Agriculture, Ecosystem and Environment, 77*, 43–52.

Ang, F., & Van Passel, S. (2010). The sustainable value approach: A clarifying and constructive comment. *Ecological Economics, 69*, 2303–2306.

Ang, F., Van Passel, S., & Mathijs, E. (2011). An aggregate resource efficiency perspective on sustainability: A Sustainable Value application to the EU-15 countries. *Ecological Economics, 71*, 99–110.

Atkinson, G. (2000). Measuring corporate sustainability. *Journal of Environmental Planning and Management, 4*, 235–252.

Azad, M. A. S., & Ancev, T. (2010). Using ecological indices to measure economic and environmental performance of irrigated agriculture. *Ecological Economics, 69*, 1731–1739.

Balana, B. B., Mathijs, E., & Muys, B. (2010). Assessing the sustainability of forest management: An application of multi-criteria decision analysis to community forests in northern Ethiopia. *Journal of Environmental Management, 91*, 1294–1304.

Bastianoni, S., Nielsen, S. N., Marchettini, N., & Jorgensen, S. E. (2005). Use of thermodynamic functions for expressing some relevant aspects of sustainability. *International Journal of Energy Research, 29*, 53–64.

Bell, S., & Morse, S. (1999). *Sustainability indicators: Measuring the immeasurable*. London: Earthscan Publications.

Bell, S., & Morse, S. (2003). *Measuring sustainability: Learning from doing*. London: Earthscan Publications.

Binder, C. R., Feola, G., & Steinberger, J. K. (2010). Considering the normative, systemic and procedural dimensions in Indicator-based sustainability assessments in agriculture. *Environmental Impact Assessment Review, 30*, 71–81.

Bossel, H. (1999). Indicators for sustainable development: Theory, method, applications (Technical report), International Institute for sustainable development.

Coelli, T., Lauwers, L., & Van Huylenbroeck, G. (2007). Environmental efficiency measurement and the materials balance condition. *Journal of Productivity Analysis, 28*, 3–12.

Cornelissen, A., van den Berg, J., Kroops, W., Grossman, M., & Udo, H. (2001). Assessment of the contribution of sustainability indicators to sustainable development: A novel approach using fuzzy set theory. *Agriculture, Ecosystems & Environment, 86*, 173–185.

Costanza, R. (2000). The dynamics of the ecological footprint concept. *Ecological Economics, 32*, 341–345.

Costanza, R., d'Arge, R., de Groot, R., Farber, S., Grasso, M., Hannon, B., Limburg, K., Naeem, S., O'Neill, R. V., Paruelo, J., Raskin, R. G., Sutton, P., & van den Belt, M. (1997). The value of the world's ecosystem services and natural capital. *Nature, 387*, 253–260.

Daly, H., & Cobb, J. (1989). *For the common good: Redirecting the economy towards community, the environment, and a sustainable future*. Boston: Beacon.

Dantsis, T., Douma, C., Giourga, C., Loumou, A., & Polychronaki, E. A. (2010). A methodological approach to assess and compare the sustainability level of agricultural plant production systems. *Ecological Indicators, 10*, 256–263.

de Haan, M. (2004). Accounting for goods and for bads: Measuring environmental pressure in a national accounts framework. Ph.D. thesis, Universiteit Twente.

De Koeijer, T., Wossink, G., Struik, P., & Renkema, J. (2002). Measuring agricultural sustainability in terms of efficiency: The case of Dutch sugar beet growers. *Journal of Environmental Management, 66*, 9–17.

De Mey, K., D'Haene, K., Marchand, F., Meul, M., & Lauwers, L. (2011). Learning through stakeholder involvement in the implementation of MOTIFS, an integrated assessment model for sustainable farming in Flanders. *International Journal of Agricultural Sustainability, 9*, 350–363.

Diaz-Balteiro, L., & Romero, C. (2004). In search of a natural systems sustainability index. *Ecological Economics, 49*, 401–405.

Ekins, P., & Simon, S. (1999). The sustainability gap: A practical indicator of sustainability in the framework of national accounts. *International Journal of Sustainable Development, 2*, 32–58.

Engstrom, R., Nilsson, M., & Finnveden, G. (2008). Which environmental problems get policy attention? Examining energy and agricultural policies in Sweden. *Environmental Impact Assessment Review, 28*, 241–255.

European Commission. (2005). Measuring progress towards a more sustainable Europe: Sustainable development indicators for the European Union (Technical report). Luxembourg: Eurostat.

Ewert, F., van Ittursum, M. K., Bezlepkina, I., Therond, O., Andersen, E., Belhouchette, H., Bockstaller, C., Brouwer, F., Heckelei, T., Janssen, S., Knapen, R., Kuiper, M., Louhichi, K., Olsson, J. A., Turpin, N., Wery, J., Wien, J. E., & Wolf, J. (2009). A methodology for enhanced flexibility of integrated assessment in agriculture. *Environmental Science & Technology, 12*, 546–561.

Farrell, A., & Hart, M. (1998). What does sustainability really mean? The search for useful indicators. *Environment, 40*, 4–9.

Figge, F., & Hahn, T. (2004). Sustainable value added – Measuring corporate contributions to sustainability beyond eco-efficiency. *Ecological Economics, 48*, 173–187.

Figge, F., & Hahn, T. (2005). The cost of sustainability capital and the creation of sustainable value by companies. *Journal of Industrial Ecology, 9*, 47–58.

Figge, F., & Hahn, T. (2009). Not measuring sustainable value at all: A response to Kuosmanen and Kuosmanen. *Ecological Economics, 69*, 244–249.

Gasparatos, A. (2010). Embedded value systems in sustainability assessment tools and their implications. *Journal of Environmental Management, 91*, 1613–1622.

Gasparatos, A., El-Haram, M., & Horner, M. (2008). A critical review of reductionist approaches for assessing the progress towards sustainability. *Environmental Impact Assessment Review, 28*, 286–311.

Gibson, C. C., Ostrom, E., & Ahn, T. K. (2000). The concept of scale and the human dimensions of global change: A survey. *Ecological Economics, 32*, 217–239.

Girardin, P., Bockstaller, C., & van der Werf, H. (2000). Assessment of potential impacts of agricultural practices on the environment: The AGRO*ECO method. *Environmental Impact Assessment Review, 20*, 227–239.

Gomez-Limon, J., & Sanchez-Fernandez, G. (2010). Empirical evaluation of agricultural sustainability using composite indicators. *Ecological Economics, 69*, 1062–1075.

Hacking, T., & Guthrie, P. (2008). A framework for clarifying the meaning of triple bottom line, integrated and sustainability assessment. *Environmental Impact Assessment Review, 28*, 73–89.

Hahn, T., Figge, F., & Barkemeyer, R. (2007). Sustainable value creation among companies in the manufacturing sector. *International Journal of Environmental Technology and Management, 7*, 496–512.

Hahn, T., Figge, F., Barkemeyer, R., & Liesen, A. (2009). Sustainable value in automobile manufacturing, report, 2nd edn. Sustainable Value Research Ltd http://www.sustainablevalue.com/downloads/sustainablevalueinautomobilemanufacturing.pdf.

Hahn, T., Figge, F., Liesen, A., & Barkemeyer, R. (2010). Opportunity cost based analysis of corporate eco-efficiency: A methodology and its application to the CO2-efficiency of German companies. *Journal of Environmental Management, 91*, 1997–2007.

Häni, F., Braga, F., Stämpfli, A., Keller, T., Ficher, M., & Porsche, H. (2003). RISE: A tool for holistic sustainability assessment at the farm level. *International Food Agribusiness Management Review, 6*, 78–90.

Hanley, N., Moffatt, I., Faichney, R., & Wilson, M. (1999). Measuring sustainability: A time series of alternative indicators for Scotland. *Ecological Economics, 28*, 55–73.

Hoang, V. N., & Rao, D. S. P. (2010). Measuring and decomposing sustainable efficiency in agricultural production: A cumulative exergy balance approach. *Ecological Economics, 69*, 1765–1776.

Jollands, N., Lermit, J., & Patterson, M. (2004). Aggregate eco-efficiency indices for New Zealand – A principal component analysis. *Journal of Environmental Management, 73*, 293–305.

Kondyli, J. (2010). Measurement and evaluation of sustainable development a composite indicator for the islands of the North Aegean region, Greece. *Environmental Impact Assessment Review, 30*, 347–356.

Kuosmanen, T., & Kuosmanen, N. (2009). How not to measure sustainable value (and how one might). *Ecological Economics, 69*, 235–243.

Langeveld, J. W. A., Verhagen, A., Neeteson, J. J., van Keulen, H., Conijn, J. G., Schils, R. L. M., & Oenema, J. (2007). Evaluating farm performance using agri-environmental indicators: Recent experiences for nitrogen management in The Netherlands. *Journal of Environmental Management, 82*, 363–376.

Lenz, R., Malkina-Pykh, I. G., & Pykh, Y. (2000). Introduction and overview. *Ecological Modelling, 130*, 1–11.

Lewis, K. A., & Bardon, K. S. A. (1998). Computer-based informal environmental management system for agriculture. *Environmental Modelling & Software, 13*, 123–137.

Lopez-Ridauro, S., Masera, O., & Astier, M. (2002). Evaluating the sustainability of complex socio-environmental systems. the MEMSIS framework. *Ecological Indicators, 2*, 135–148.

Marchand, F., De Mey, K., Debruyne, L., D'Haene, K., Meul, M., Lauwers, L. (2010). From individual behaviour to social learning: Start of a participatory process towards sustainable agriculture. *Proceedings of the 9th European IFSA Symposium, Building sustainable rural futures*, (pp. 670–682).

Meul, M., Nevens, F., Reheul, D., & Hofman, G. (2007). Energy use efficiency of specialized dairy, arable and pig farms in Flanders. *Agriculture, Ecosystems & Environment, 119*, 135–144.

Meul, M., Van Passel, S., Nevens, F., Dessein, J., Rogge, E., Mulier, A., & Van Hauwermeiren, A. (2008). MOTIFS: A monitoring tool for integrated farm sustainability. *Agronomy for Sustainable Development, 28*, 321–323.

Meul, M., Nevens, F., & Reheul, D. (2009). Validating sustainability indicators: Focus on ecological aspects of Flemish dairy farms. *Ecological Indicators, 9*, 284–295.

Moffatt, I. (2000). Ecological footprints and sustainable development. *Ecological Economics, 32*, 359–362.

Neumayer, E. (2003). *Weak versus strong sustainability: Exploring the limits of two opposing paradigms*. Cheltenham: Edward Elgar.

Nevens, F., Verbruggen, I., Reheul, D., & Hofman, G. (2006). Farm gate nitrogen surpluses and nitrogen use efficiency of specialized dairy farms in Flanders: Evolution and future goals. *Agricultural Systems, 88*, 142–155.

Nevens, F., Dessein, J., Meul, M., Rogge, E., Verbruggen, I., Mulier, A., Van Passel, S., Lepoutre, J., & Hongenaert, M. (2008). On tomorrow's grounds; development of a vision on Flemish agriculture in 2030. *Journal of Cleaner Production, 16*, 1062–1070.

Nykvist, B., & Nilsson, M. (2009). Are impact assessment procedures actually promoting sustainable development? Institutional perspectives on barriers and opportunities found in the Swedish committee system. *Environmental Impact Assessment Review, 29*, 15–24.

Odum, H. T. (1996). *Environmental accounting: Emergy and environmental decision making*. New York: Wiley.

OECD. (2006). *OECD factbook 2006, economic, environmental and social statistics*. Paris: Organisation for Economic Co-operation and Development.

Pacini, C., Giesen, G., Wossink, A., Omodei-Zorini, L., & Huirne, R. (2004). The EU's Agenda 2000 reform and the sustainability of organic farming in Tuscany: Ecological-economic modelling at field and farm level. *Agricultural Systems, 80*, 171–197.

Pannell, D. J., & Glenn, N. A. (2000). A framework for the economic evaluation and selection of sustainability indicators in agriculture. *Ecological Economics, 33*, 135–149.

Patterson, M. (2006). Selecting headline indicators for tracking progress to sustainability in a nation state. In P. Lawn (Ed.), *Sustainable development indicators in ecological economics* (pp. 421–448). Cheltenham/Northamptom: Edward Elgar.

Pearce, D. W., & Atkinson, G. D. (1993). Capital theory and the measurement of sustainable development: An indicator of weak sustainability. *Ecological Economics, 8*, 103–108.

Poppe, J., Annadale, D., & Morison-Saunders, A. (2004). Conceptualising sustainability assessment. *Environmental Impact Assessment Review, 24*, 595–616.

Reinhard, S., Lovell, C., & Thijssen, G. (2000). Environmental efficiency with multiple environmentally detrimental variables; estimated with SFA and DEA. *European Journal of Operational Research, 121*, 287–303.

Rigby, D., Woodhouse, P., Young, T., & Burton, M. (2001). Constructing a farm level indicator of sustainable agricultural practice. *Ecological Economics, 39*, 463–478.

Rodrigues, G. S., Rodrigues, I. A., Buschinelli, C. C. A., & de Barros, I. (2010). Integrated farm sustainability assessment for the environmental management of rural activities. *Environmental Impact Assessment Review, 30*, 229–239.

Saisana, M., Tarantola, S., Schulze, N., Cherchye, L., Moese, W., Puyenbroeck, T. V. (2005). State-of-the-art report on composite indicators for the knowledge-based economy (Technical report), KEI, Knowledge Economy Indicators project.

Sands, G. R., & Podmore, T. H. (2000). A generalized environmental sustainability index for agricultural systems. *Agriculture, Ecosystems & Environment, 79*, 29–41.

Sauvenier, X., Valckx, J., Cauwenbergh, N. V., Wauters, E., Bachev, H., Biala, K., Bielders, C., Brouckaert, V., Garcia-Cidad, V., Goyens, S., Hermy, M., Mathijs, E., Muys, B., Vanclooster, M., Peeters, A. (2005). Framework for assessing sustainability levels in Belgian agricultural systems – SAFE (Technical report), Scientific Support Plan for a Sustainable Development Policy.

Schultink, G. (2000). Critical environmental indicators: Performance indices and assessment models for sustainable rural development planning. *Ecological Modelling, 130*, 47–58.

Sheate, W. R., & Partidario, M. R. (2010). Strategic approaches and assessment techniques – Potential for knowledge brokerage towards sustainability. *Environmental Impact Assessment Review, 30*, 278–288.

Shields, D., Solar, S., & Martin, W. (2002). The role of values and objectives in communicating indicators of sustainability. *Ecological Indicators, 2*, 149–160.

Smith, C. S., McDonald, G. T., & Thwaites, R. N. (2000). TIM: Assessing the sustainability of agricultural land management. *Journal of Environmental Management, 60*, 267–288.

Stoorvogel, J. J., Antle, J. M., Crissman, C. C., & Bowen, W. (2004). The tradeoff analysis model: Integrated bio-physical and economic modeling of agricultural production systems. *Agricultural Systems, 80*, 43–66.

ten Berge, H., van Ittersum, M., Rossing, W., van de Ven, G., Schans, J., & van de Sanden, P. (2000). Farming options for the Netherlands explored by multi-objective modeling. *European Journal of Agronomy, 13*, 263–277.

United Nations. (2001). Indicators of sustainable development: Guidelines and methodologies (Report), United Nations.

van Calker, K., Berentsen, P., de Boer, I., Giesen, G., & Huirne, R. (2004). An LP-model to analyze economic and ecological sustainability on Dutch dairy farms: Model presentation and application for experimental farm de Marke. *Agricultural Systems, 82*, 139–160.

van Calker, K., Berentsen, P., Romero, C., Giesen, G., & Huirne, R. (2006). Development and application of a multi-attribute sustainability function for Dutch dairy farming systems. *Ecological Economics, 57*, 640–658.

Van Cauwenbergh, N., Bielders, K. B. C., Brouckaert, V., Franchois, L., Garcia Cidad, V., Hermy, M., Mathijs, E., Muys, B., Reijnders, J., Sauvenier, X., Valckx, J., Vanclooster, M., Van der Veken, B., Wauters, E., & Peeters, A. (2007). SAFE – A hierarchical framework for assessing the sustainability of agricultural systems. *Agriculture, Ecosystems & Environment, 120*, 229–242.
Van Passel, S., Nevens, F., Mathijs, E., & Van Huylenbroeck, G. (2007). Measuring farm sustainability and explaining differences in sustainable efficiency. *Ecological Economics, 62*, 149–161.
Van Passel, S., Van Huylenbroeck, G., Lauwers, L., & Mathijs, E. (2009). Sustainable value assessment of farms using frontier efficiency benchmarks. *Journal of Environmental Management, 90*, 3057–3069.
von Wirén-Lehr, S. (2001). Sustainability in agriculture – An evaluation of principal goal-oriented concepts to close the gap between theory and practice. *Agriculture, Ecosystems & Environment, 84*, 115–129.
Wackernagel, M., & Rees, W. E. (1997). Perceptual and structural barriers to investing in natural capital: Economics from an ecological footprint perspective'. *Ecological Economics, 20*, 3–24.

Part III
Application of Methodologies for Sustainability Assessment of Farming Systems in Real Contexts

Chapter 7
Learning and Experimentation Strategy: Outline of a Method to Develop Sustainable Livestock Production Systems

Boelie Elzen and Sierk F. Spoelstra

Abstract Over the past decade, the Dutch government has increasingly emphasised the need for integral solutions for sustainability problems in the livestock production sector. This led to the adoption of research approaches in line with transition management and system innovation that had been developed in other domains. In 2008, the government set further policy targets of 5 and 100% sustainable livestock production at the farm level for 2011 and 2023, respectively. Policy measures included stimulation of sector initiatives for sustainable agriculture (sectoral innovation agendas) and demand for projects with a focus on system innovation. Two broad approaches may contribute to the realisation of these targets, notably top-down and bottom-up. Currently, the links between the bottom-up and the top-down processes are relatively weak. As both may contribute to a system innovation, a major challenge is to make a fruitful combination between the two approaches. To this end, we have developed what we call a 'learning and experimentation strategy' (LES) that we will elaborate in this chapter.

Keywords Livestock production systems • Top-down • Bottom-up • Learning and experimentation strategy • Sustainability solutions

7.1 Introduction

During the second half of the twentieth century, livestock production in the Netherlands evolved in a close alignment between politics, policy and sector representatives. The main focus was on increasing production efficiency with a strong

B. Elzen (✉) • S.F. Spoelstra
Wageningen UR Livestock Research, P.O. box 65, 8200 AB
Lelystad, The Netherlands
e-mail: b.elzen@utwente.nl; sierk.spoelstra@wur.nl

A.A. Marta-Costa and E. Silva (eds.), *Methods and Procedures for Building Sustainable Farming Systems: Application in the European Context*,
DOI 10.1007/978-94-007-5003-6_7,

orientation towards export. Gradually, this modernisation process became criticised for its negative side effects. Early criticism emphasised the dangers of chemical pest and weed control, emission of malodours from livestock units and mineral surpluses. Later, emphasis shifted to impaired animal welfare and contagious and zoonotic animal diseases, especially after outbreaks of a variety of epidemic animal diseases in the past decade, including classical swine fever, foot and mouth disease, avian influenza and BSE. Recently, criticism centred on contribution of livestock production to climate change and to excessive claims on natural resources of food production.

Governmental policies aimed to solve or mitigate the problems by stimulating research, subsidy programmes and regulatory actions. In most cases, these measures led to reducing the specific problem by technical means and regulations for the livestock production system. Thus, the agricultural system that had emerged during the first modernisation (Beck 1986) met the first attempts of reflexive modernisation. The latter, however, also used various thoughts and approaches (hard and soft institutions) rooted in modernity. Thus, the actors involved on the one hand continued to increase production efficiency and on the other tried to fine-tune inputs (of nutrients, agrochemicals, manure, etc.) to societal needs.

Since the mid-1990s, the search for integral solutions gradually received attention, which led to governmental policy partially adopting research approaches in line with transition management and system innovation that had been developed in other domains. In 2008, the Dutch government set specific policy targets of 5 and 100% sustainable livestock production at the farm level for 2011 and 2023, respectively (LNV 2008). Policy measures included stimulation of sector initiatives for sustainable agriculture (sectoral 'innovation agendas'), demand for projects with a focus on system innovation and societal design and subsidy instruments for agricultural entrepreneurs and integral research.

To meet the challenges in the livestock production sector, two broad approaches evolved: top-down and bottom-up. Top-down approaches are typically research led and often start with the formulation of visions of future livestock production systems. These include redesign of primary production (Bos and Grin 2008), inclusion of new functions in primary production, vertical integration in the supply chain and combining functions of different agricultural activities in agro-production parks (Grin and Van Staveren 2007). The underpinning of the sustainability claim of such visions varies from expert analysis only, results of extensive stakeholder consultation to deliberate co-design by scientific experts and stakeholders.

At the same time, a broad variety of bottom-up initiatives are taken by farmers who develop and try out new approaches to meet the challenges as they see them. Most of these initiatives are not guided by broad future visions and focus on specific aspects.

Currently, the links between the bottom-up and the top-down processes are relatively weak. From the top-down perspective, the bottom-up initiatives are even considered risky since they typically address a relatively small problem within the current system and might solidify the system rather than opening it up, whereas the top-down approaches explicitly seek to change the system at large.

However, a system innovation can never be 'organised from above'. It needs to make use of the 'innovative energies' within the existing livestock production sector, i.e. lessons learned in the bottom-up process. A major challenge is to make a fruitful combination between the top-down and bottom-up processes. It is this challenge that we will address in this chapter.

Much research has been done on top-down approaches like strategic niche management (Hoogma et al. 2002; Schot and Geels 2008) and transition management (Rotmans 2003; Loorbach 2007). For this reason, we will focus on the bottom-up processes in this chapter but within the overall ambition of combining this perspective with top-down approaches. We will present a tentative framework to assess bottom-up initiatives as well as top-down projects on their potential to contribute to system innovation. This framework serves as a tool in a broad learning and experimentation strategy in which the lessons from top-down and bottom-up are combined. We are currently (summer 2012) testing this framework in various sectors, and on the basis of this, we will modify and elaborate it for wider applicability.

7.2 The Dynamics of System Innovation

The central issue in this chapter is how learning and experimentation in projects may contribute to system innovation. The traditional model sees innovation as a diffusion process: via innovators, early adopters, early majority, late majority and eventually laggards (Rogers 1962). Also, system innovations have been portrayed as a sort of diffusion process, distinguishing the following phases: pre-development, take-off, acceleration and stabilisation (Rotmans 2003).

Although extensive later work has shown that these diffusion models are oversimplistic, they are still widely held valid in policy arenas and also in scientific circles (e.g. Gielen and Zaalmink 2003). Policy makers, after a successful project, immediately tend to pose the question 'And now, how do we scale up?' The so-called multilevel perspective (MLP; Rip and Kemp 1998; Geels 2002) provides a more dynamic view on innovation. The core of the MLP is that system innovations are shaped by interaction between three levels: the socio-technical landscape, the socio-technical regimes and technological niches (Fig. 7.1). Socio-technical systems are located at the meso-level of *socio-technical* regimes. These regimes indicate a set of shared rules that guide and constrain the actors within a production and consumption system in how they try to tackle various challenges they encounter. This typically leads to evolutionary patterns of innovation. The *socio-technical landscape* is an exogenous environment of factors with a broader societal relevance like the need to reduce CO_2 emissions. *Technological niches* are the breeding ground for radical innovations that initially poorly fit the regime.

In the MLP dynamic, system innovations develop as follows. A novelty emerges in a local practice and becomes part of a niche when a network of actors is formed that share certain expectations about the future success of the novelty and are willing to fund and work on further development. Niches may emerge and develop

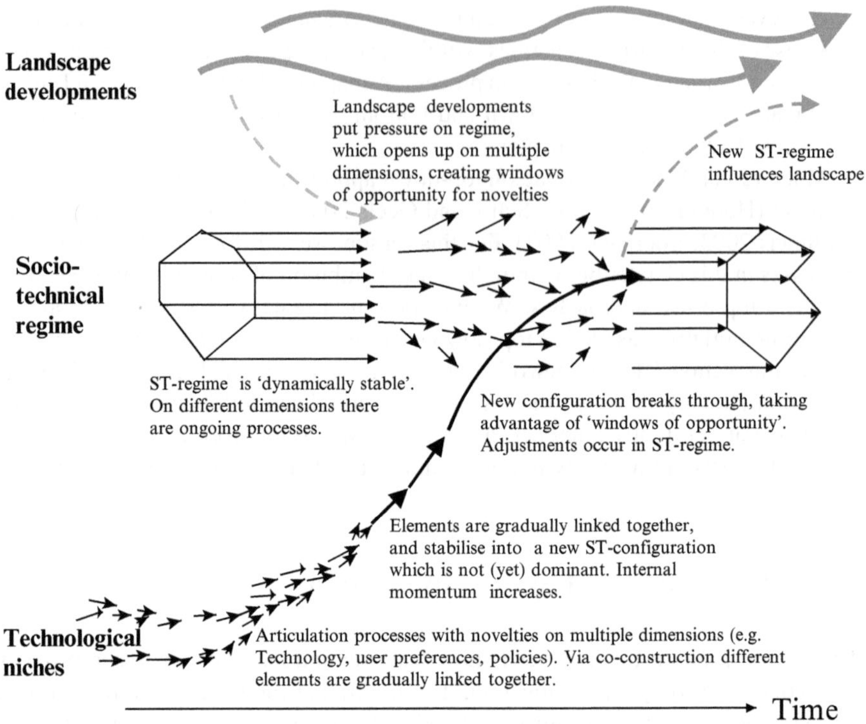

Fig. 7.1 A dynamic multilevel perspective on system innovation (Geels and Schot 2007)

partly in response to pressure and serious problems in an existing regime which can be either internal to the regime itself (such as animal welfare in industrial animal production) or come from the socio-technical landscape (e.g. the pressure to curb CO_2 emissions which affects more than just the animal production sector). The further success of niche formation is on the one hand linked to processes within the niche (micro-level) and on the other hand to developments at the level of the existing regime (meso-level) and the socio-technical landscape (macro-level). Supported by actors willing to invest in the new concept (industries, R&D organisations, government) and initially protected from competition at the market place (e.g. through subsidies), the technology is improved within the niche, broader networks are formed around it and more is learned about directions for improvement and functions it may fulfil.

After some level of improvement of the technology and after learning more about its potential, it may find its way in specific market applications, often typical segments that exploit new functional characteristics of the technology and focus less on cost structures (e.g. organic food). Through further improvement, increasing reliability and cumulated experiences and learning about functionalities and potential applications, the technology can spread to other market niches and/or trigger expansion of market niches. Processes of rule formation also play an important role, such as the development of standards and regulations for the technology, and processes that reduce the mismatch of the emerging technology with the rules of the dominant regime. As it starts to compete on or with main markets, the novelty may transform or substitute the existing regime and thus trigger a system innovation process.

This perspective allows for a very dynamic view on innovation processes as its application to a variety of historical cases has shown. These studies, however, tend to focus on the vicissitudes of a specific alternative technology to an existing system (e.g. sailing ships replacing steamships; Geel, 2002) although the new technology does not simply diffuse but changes in the process. This works fine for retrospective studies, but it is problematic to use as a heuristic in a 'learning and experimentation strategy' seeking to contribute to system innovation. We do not know which alternative development will play a key role in the development towards a sustainable livestock sector. We need to acknowledge that 'innovation in action' is much messier than retrospective historical studies portray it (see, e.g. Elzen et al., forthcoming).

7.3 Portfolio of Promises

In the MLP, niches are the locus to learn about and further develop novelties. A niche consists of a variety of projects that share a technical nucleus, e.g. electric propulsion for cars (Hoogma et al. 2002). Using the niche concept in a sector like animal production, however, is problematic because innovative projects and practice initiatives are very diverse. For instance, they may relate to new types of animal food, new manure collection technologies, new husbandry systems, etc. Learning between these initiatives is often minimal, and, therefore, they do not fit the definition of a niche in the MLP.

To address such innovations, we will use the term 'promise'. The term promise expresses that each of these novelties has attractive sides from a certain sustainability perspective (e.g. lower CO_2 emissions), but it has also problematic (e.g. more expensive) or unknown sides. Initially, a promise may just be an idea or a concept, explored in a single project. After a certain period of time, more projects may be started in connection with the promise. When these projects start exchanging information, the promise may thus develop into a niche.

Historical cases show that system innovations are not the result of the 'massive diffusion' of a new technology but a lengthy process of combining and recombining 'partial innovations'. This implies that, to induce or stimulate system innovations, attention should not go to a single novelty (or promise) but to range of novelties that we call the 'portfolio of promises'. In a project seeking to develop new 'integrally sustainable' husbandry systems for dairy cows ('Kracht van Koeien' (Cow Power); cf. Bos 2009), we distinguish about a dozen such promises, including separate collection and processing of manure and urine, minimum space of 360 m^2 per cow throughout the year, cheap but sustainable roofed shelters (rather than a closed barn), etc.

For each of these promises, a process of learning and experimentation is needed to find out in practice how the problematic sides may be solved and to explore whether new sustainability problems are created. For an individual promise, even if it does not (yet) constitute a niche, the approach of strategic niche management (SNM) provides valuable suggestions on how to do this (Hoogma et al. 2002; Schot and Geels 2008). But SNM looks at the level of a single novelty and not at the portfolio level, i.e. across a variety of niches in MLP terms. To make a more encompassing contribution to system innovation, we need a learning and experimentation strategy that works at two levels, at the level of individual promises and at the level of the portfolio of promises.

- The individual promise level: Because we are not only looking at technical innovations but also at new practices, new meanings, etc., it is important to make various stakeholders, to whom the experiment may be relevant, part of the network exploring it (e.g. the 'roofed shelter network' in the Cow Power project mentioned above). Because a wide variety of 'partial innovations' will be required for a system innovation, a large number of such networks will be required for a long period (as system innovation tends to be a lengthy process).
- The portfolio level: Because a system innovation will result from a process of combining and recombining partial innovations, it is important to analyse how various promises might be linked to create a full system that is more sustainable than the existing one. Such an analysis at the portfolio level (the 'portfolio integration') may result in starting new experiments with linked promises (thus creating a new, more encompassing promise) or in giving feedback to ongoing experiments to include certain aspects based on the portfolio integration. Because a variety of promise networks need to be running for a longer period, this portfolio integration should be a more or less continuous activity.

This combination of learning and experimentation at two levels we call the 'learning and experimentation strategy' (LES). It can be seen as an extension of SNM in two directions: (1) It addresses promises before they constitute a niche, and (2) it looks across a range of promises (or multiple niches in SNM terms). In the next section, we will show that LES has a further extension compared to SNM (as well as to transition management) by incorporating 'top-down' as well as 'bottom-up' initiatives.

7.4 Two Complementary LES Approaches: Top-down and Bottom-up

Historical system innovations have rarely been planned, and they usually developed solely out of bottom-up processes. The idea to deliberately evoke system innovations for societal goals (like sustainable animal production) is relatively new. In top-down approaches like SNM and transition management (TMgt), organised projects are crucial in achieving this. Organising projects, however, does not imply that the bottom-up dynamic has been halted, but this is ignored in SNM and TMgt. A 'complete' approach to evoke system innovations should combine the top-down and the bottom-up processes. We will discuss each of these below.

7.4.1 Top-down

Generally, top-down approaches are research led and start with the exploration of possible sustainable futures (Hirsch Hadorn et al. 2008). The nature of such explorations varies widely and could be based on extrapolation of trends, scenarios, dynamic modelling, elaborating visions and actions of co-design or *ad hoc* methods to define requirements for a future system without the problems of the existing one. Future explorations serve functions like giving directions to short-term actions, a certain loosening up from today's preoccupations and achieving opening up and congruency among stakeholders about a future orientation. Smith et al. (2005) distinguish the following functions of a future exploration or vision-building exercise:

- Mapping a 'possibility space': Visions identify a realm of plausible alternatives for conceiving of socio-technical functions and for the means of providing for them.
- A heuristic: Visions act as problem-defining tools by pointing to technical, institutional and behavioural problems that need to be resolved.
- A stable frame for target setting and monitoring progress: Visions stabilise technical and other innovative activity by serving as a common reference point for actors collaborating on its realisation.
- A metaphor for building actor networks: Visions specify relevant actors (by inclusion and exclusion), acting as symbols that bind together communities of interest and of practice.
- A narrative for focusing capital and other resources: Visions become an emblem that is employed in the marshalling of resources from outside an incipient regime's core membership (see also Rotmans 2003; Loorbach 2007; Berkhout et al. 2004; Brown et al. 2000).

In the Netherlands, the approach of sustainable technological development (STD; Weaver et al. 2000) has gained considerable attention. It starts by constructing visions of a desirable future and then uses a method called backcasting to define short-term actions. The backcasting is carried out in interaction with stakeholders (Quist 2007). The approach of transition management follows a comparable methodology (Rotmans 2003). Here, a 'basket of visions' is developed with a variety of stakeholders which are also 'translated back' into concrete projects in the near term.

In our view, these top-down approaches take too much of a planning approach towards developing the future. Innovation in practice is a very messy process in which a wide variety of stakeholders are active, and one of the challenges is to use the 'innovative energy' that is already there. To achieve this, we have been involved in vision-building exercises with sectoral stakeholders for various livestock sectors, including laying hens, pigs and dairy cows. Most often, the visions take the form of a report or brochure giving general 'contours' of more sustainable husbandry systems for a sector along with concrete suggestions for various 'subsystems' (the 'promises'). Via various communication outlets, we make these images widely known in the sector and invite concrete farmers to try and implement various aspects of it on their own farm. For laying hens, this resulted in a new system by the name of Roundel that is currently experimented with by concrete farmers (Groot Koerkamp and Bos 2008; Klerkx et al. 2009). For dairy cows, visions of four sustainable new systems have been launched early 2009 (Bos 2009) and since we have been frequently approached by farmers who want to try out aspects of it. One of the promises now tried out by various farmers is new floors for cow houses. New floors could make contributions to sustainability aspects including animal welfare and reduction of emissions (esp. ammonia by early separation of manure and urine). In 2010, a project on new husbandry systems for fattening pigs was concluded,[1] and its results were taken as a point of departure for an innovation programme for the sector.

7.4.2 Bottom-up

The initiatives that are inspired by these visions can be seen as part of a 'top-down' dynamic which is fed by the explicit goal to develop 'integrally sustainable' husbandry systems. But we have to be modest because most of the innovative activity in a sector develops bottom-up, and much of this is *not* (or hardly) influenced by global sustainability visions. Since these 'bottom-up' initiatives outweigh the top-down initiatives by far in numbers, this begs the question whether and, if so, how the bottom-up initiatives could also be incorporated in a learning and experimentation strategy.

Let us take a closer look at this bottom-up process, i.e. the ongoing process of innovation in the animal production sector that takes place for a variety of reasons. This does not mean that such actions are not guided by visions. They usually are, but these visions tend to be of a more local nature or address a specific dimension of sustainability (rather than the 'integrally sustainable' visions in the top-down approaches).

We can take two different views at the agricultural (including animal production) sector. In the first, agriculture basically refers to the primary production at the farm with the goal of producing all sorts of food products (called 'conventional agriculture'). By far, the largest volume of agricultural products is produced in a rather uniform fashion. Important characteristics of this system are cost price competitiveness and production for international food corporations (cf. Van der Ploeg 2008). Innovation focuses on this competitiveness. Other directions for innovation are neglected, and the embedding of agriculture in the existing system is considered self-evident. Visions of change are confined to the farm level or the desire that the food processing industry take the lead (cf. the 'Innovation Agenda' for the pig husbandry sector in the Netherlands). In such a view, local innovative initiatives are hardly relevant. They may lead to nice niche products, but they will hardly contribute to sustainable development.

In the second view, by contrast, the multitude of local initiatives is seen as a source of potential change and inspiration. These initiatives are not only seen as an

[1] http://www.duurzameveehouderij.wur.nl/UK/projects/porkopportunities/

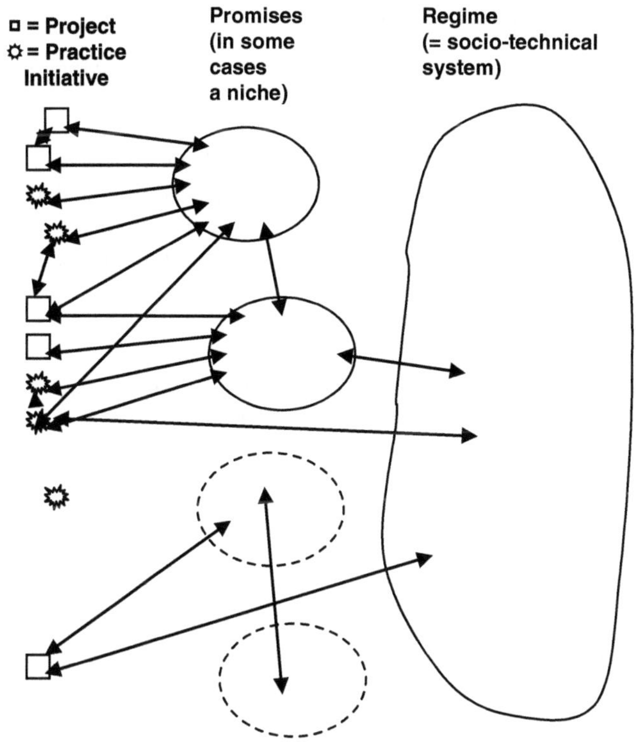

Fig. 7.2 LES concepts in the multilevel dynamic

effort to innovate at the farm level, but they are inseparable from their institutional embedding. Roep et al. (2003) refer to this process in the agricultural sector as 'technological-institutional' design which is connected to what they call effective reformism. Their basic idea is that especially in the agricultural sector, the initiatives from farmers typically aim at simultaneously realising technical change as well as creating a new institutional environment (new routines and links with various stakeholders, including advisors, supplier and processing corporations, public authorities, the general public, etc.). In this process, the expectations of farmers as well as the other stakeholders change. Thus, such initiatives may form the 'seeds of transition' (Wiskerke and Van der Ploeg 2004; see also Roep and Wiskerke 2006) although they are not guided by 'integral sustainability' visions.

This means that such bottom-up initiatives are certainly relevant for a learning and experimentation strategy for sustainability. LES should analyse the contributions of such bottom-up initiatives as well as top-down organised projects concerning the contribution they make or can make to the development of individual promises as well as the whole portfolio.

This is captured in Fig. 7.2 which gives a representation of the multilevel dynamic, focusing on the relationships between projects, practice initiatives, promises and the regime.

Some explanatory remarks concerning the figure are as follows:

- Two-way arrows are used to indicate that influences may go both ways.
- Projects and initiatives may contribute to more than one promise.
- Some of the promises have been dashed, indicating they are (still) conceptual ideas that are not or hardly supported by a network. One of these is not supported by any project or initiative, indicating it is still just a conceptual idea.
- Projects and practice initiatives may influence each other directly.
- Projects and practice initiatives may also have an influence on the regime directly.
- Promises may also influence the regime.
- Promises may influence one another.
- One isolated initiative is not connected to any promise as an example of many such initiatives that do not fit the portfolio of promises.

7.5 Assessing Promises

Farmers apply innovations for a variety of reasons. There may be thousands of such initiatives, some of which may be inspired for sustainability reasons, while many others are motivated otherwise. This begs the question how to assess which initiatives might make a contribution to sustainable development. This is not simply a matter of listening to the farmers' motivations as historical studies show that later developments may go in directions very different from what the initiators intended or aspired to.

We can approach this issue in various ways. First, we may ask the question 'Which initiatives are sustainable?' This may sound like an over-simplistic question, but it is one that the current political situation in the Netherlands (as well as in many other countries) confronts us with. A 2008 white paper from the minister of agriculture states that by 2011, 5% of the Dutch husbandry systems should be sustainable (LNV 2008). Therefore, the ministry needed criteria that allowed counting whether the target had been met. In the Netherlands, such criteria have been and are being developed in the form of sustainability indexes for various agricultural subsectors. These indexes provide criteria that are assessed in a quality assurance scheme (cf. www.smk.nl).

The second approach in assessing bottom-up initiatives is to see them as part of an ongoing process. The question then becomes 'Which initiatives have a potential to contribute to sustainable animal production?' This requires a broader set of assessment criteria such as the presence of a broader vision on sustainability, institutional embedding and change, risk insurance for individual farmers, room to learn and experiment, a potential to apply the innovation in a commercial setting eventually (e.g. via initial financial support), etc. Such criteria are more qualitative than under the first approach and more open for debate.

As a third approach, the question may be reversed. 'How can we use these initiatives to learn about possibilities for sustainable animal production?' The initiatives are then seen as learning experiments to render knowledge on barriers and chances for sustainable development. Thus, they can be made part of the 'portfolio of promises' within LES. This requires a process of continuous monitoring of which innovations are explored in the animal production practice and assess the relevance of the locally learned lessons within the broader portfolio.

An important aspect of this third approach is that bottom-up initiatives (esp. when analysed in combination with top-down projects) can be used to learn about uncertainties. Some of the main uncertainties are (1) whether the envisaged innovation compares favourably to existing practices, (2) whether the innovation produces new unforeseen risks when applied during a longer period and on a larger scale and (3) how the innovation potentially compares with other competing solutions. Furthermore, in connection with the overall goal of system innovation, a major uncertainty is whether the combination of top-down and bottom-up learning and experimentation might eventually lead to weakening of the existing regime and to a shift towards another socio-technical system.

In LES, we follow a combination of the second and third approaches. The points raised above imply that we need a tool to assess the various promises on their potential contribution to sustainable animal production. Tentatively, we are now using an evaluation framework in which we assess each initiative on the following dimensions:

- Sustainability gains/losses: An assessment of whether the novelty might improve sustainability on various sub-dimensions as PPP and animal welfare.
- System renewing potential: An assessment of whether the promise might help break the lock-in in the existing system.
- Risk of strengthening lock-in: The reverse of the previous point: Is there a risk that the innovation might consolidate the existing system and block further renewal for a long time ahead (e.g. because of huge investments made)?
- Give momentum to change processes: Does the novelty set things in motion that can be expected to continue for a considerable time?

For each of these main dimensions, we distinguish various sub-dimensions. We are currently (summer 2012) testing this framework by applying it to the dairy cow sector. We are exploring whether this leads to a meaningful comparison of various promises (top-down projects as well as bottom-up initiatives) and whether this serves as a good starting point for an analysis at the portfolio level. This empirical testing is likely to lead to some changes in the methodology and thus helps us to refine the learning and experimentation strategy that we seek to develop.

7.6 Challenges for Further LES Development

We are developing LES as a strategy to contribute to system innovation via a combination of learning from projects and learning from practice initiatives. To realise this ambition poses a number of challenges and raises a number of questions. Below, we list a number of aspects that need further elaboration:

- *Promises Monitor.* Various system innovation projects in the Netherlands are being monitored to optimise learning, but it is necessary to extend this to monitoring of practice initiatives. This raises various new questions, e.g.: Which of the numerous initiatives to actually follow? The evaluation framework in the previous section can be used to help make such decisions.

- *Promises Analysis and Evaluation.* The results of a variety of projects and bottom-up initiatives need to be evaluated in relation to each other. But how to do this and translate this into topics for further exploration (e.g. via new projects) still needs to be developed. The evaluation framework above provides a first stepping stone for this.
- *Portfolio Analysis and Evaluation.* The next step is to move beyond the promises level and analyse the data collected at the portfolio level. We still need to develop methods for relating data on various promises. One starting point may be to evaluate the data against the background of various visions of a sustainable new system (e.g. as developed in the Cow Power project), but this would still require new evaluation methods.
- *Portfolio Management.* In the present situation, management takes place at the level of projects and, to a lesser extent, at the level of programmes. A learning an experimentation strategy, however, would also require forms of management at the portfolio level which are currently non-existent.
- *Stakeholder Management.* A system innovation will require contributions from a variety of stakeholders which can only be realised by involving them in various activities in LES. But it is still an open question who to involve in which of the tasks above.

7.7 Conclusion

The societal and political pressure to develop more sustainable animal production systems (as well as other agricultural systems) has grown over the past decade and is not likely to go away. This will require system innovations in various sectors (such as animal production) and sub-sectors (e.g. dairy cows and pigs). Approaches like SNM and TMgt have rendered a variety of suggestions on how to use learning in series of projects to contribute to sustainable development.

These approaches, however, ignore that in the ongoing dynamic in these sectors, a large number of stakeholders are tinkering with a variety of innovations trying to solve a range of problems as they experience them. In historical system innovations such bottom-up processes were the dominant drivers for transitions. Current attempts that seek to evoke system innovations towards sustainability therefore cannot ignore this bottom-up dynamic and should make it part of their strategies.

The LES approach that we propose here does acknowledge this bottom-up dynamic. It attempts to combine the learning that takes place in bottom-up practice initiatives (often farmer led) with the more deliberate attempts at learning in planned projects that are often research led. This combination does more justice to the innovation dynamic that is actually taking place than the more narrow focus on projects by approaches like SNM and TMgt.

Combining top-down and bottom-up in LES also allows combining the strong and weak sides of each of these approaches, notably as follows:

- Top-down approaches are driven by the development of a vision (or set of visions) of an integrally sustainable new system. Thus, sustainability goals are baked into the process. The weak point is that these new visions and their constituting parts (the promises) do not fit in well with the existing system. This makes it difficult to 'anchor' (cf. Elzen et al. forthcoming) these novelties within the current system and gain practical experience. Such an anchoring, however, is required to get a transformation process going. Starting this process 'from the outside' is difficult and may trigger a lot of resistance.
- In bottom-up initiatives, such anchoring is guaranteed since the initiatives come from within the existing system. But because of this anchoring, it is difficult to take along broader sustainability issues which would require more radical steps.

In current practice (also in transition initiatives in other sectors), top-down (i.e. driven by integral sustainability visions) and bottom-up constitute separate approaches. Certain parties may be working on one approach who are hardly in touch with parties working on the other approach. Both, however, will contribute to the system innovations that are in the making. Furthermore, because each of these has its weak and its strong sides, it is important to link them in a learning and experimentation strategy, LES.

Current policies often make a distinction between improving sustainability in the short term by adapting existing systems and working on integral sustainability in the long term through system innovation. Bottom-up initiatives are primarily seen as contributing to the former which, however, constitutes a limited view. Judging such initiatives on direct sustainability criteria may indeed provide information on their potential to make short-term contributions. However, also incorporating other criteria (cf. the evaluation framework above) may reveal their potential to contribute to more integral sustainability in the long term as well. This also provides the opportunity to link learning from bottom-up initiatives to learning in various top-down-inspired projects. Subsequently, by 'zooming out' to the portfolio level, an integral analysis may generate new ideas on how linking between various promises (irrespective of whether they come from top-down or bottom-up learning) could result in identifying a 'higher level' promise as a contribution to a system innovation. Such a broader learning and experimentation strategy thus attempts to combine (1) top-down and bottom-up approaches and (2) the individual promise (which in some cases may be a niche) and the portfolio levels. Thus, it seeks to make a much more effective use of existing innovative potential in the sector than other approaches, and it is likely to make a larger contribution towards developing a sustainable livestock production sector as well as other sectors.

Acknowledgements This chapter is part of a research project supported by the Dutch Ministry of Agriculture, Nature and Food Quality (currently Ministry of Economy, Agriculture and Innovation; project BO-12.2-001-004) and the Wageningen UR KB-VI Programme 'Transitie en Innovatie' (Transition and Innovation).

References

Beck, U. (1986). *Risk society: Towards a new modernity*. London: Sage.

Berkhout, F., Smith, A., & Stirling, A. (2004). Socio-technical regimes and transition contexts. In B. Elzen, F. W. Geels, & K. Green (Eds.), *System innovation and the transition to sustainability: Theory, evidence and policy* (pp. 48–75). Cheltenham: Edward Elgar.

Bos, A. P. (2009, June). *Concepts and objects as boundary objects for sustainable animal husbandry: Anticipating regime transformations by design*. Paper presented at the 1st European Conference on Sustainability Transitions.

Bos, A. P. (Bram), & Grin, J. (2008). Doing reflexive modernization in pig husbandry: The hard work of changing the course of a river. *Science, Technology, & Human Values, 33*, 480–507.

Brown, N., Rappert, B., & Webster, A. (Eds.). (2000). *Contested futures: A sociology of prospective techno-science*. Aldershot: Ashgate.

Elzen, B., Leeuwis, C., & Van Mierlo, B. (forthcoming). Anchoring of innovations: Assessing Dutch efforts to use the greenhouse effect as an energy source.

Geels, F. W. (2002). Technological transitions as evolutionary reconfiguration processes: A multi-level perspective and a case-study. *Research Policy, 31*, 1257–1274.

Geels, F. W., & Schot, J. W. (2007). Typology of sociotechnical transition pathways. *Research Policy, 36*, 399–417.

Gielen, P., & Zaalmink, W. (2003) *Hoe komen gewone boeren tot verandering?* (Report 18). Wageningen: Agro Management Tools.

Grin, J., & Van Staveren, A. (2007). *Werken aan systeeminnovaties. Lessen uit de praktijk van InnovatieNetwerk*. Assen: Van Gorcum.

Groot Koerkamp, P. W. G., & Bos, A. P. (2008). Designing complex and sustainable agricultural production systems: An integrated and reflexive approach for the case of table egg production in the Netherlands. *New Jersey Academy of Sciences, 55*, 113–138.

Hirsch Hadorn, G., Hoffmann-Riem, H., Biber-Klemm, S., Gossenbacher-Mansuy, W., Joye, D., Pohl, C., Wiesmann, U., & Zemp, E. (2008). *Handbook of transdisciplinary research*. Berlin: Springer.

Hoogma, R., Kemp, R., Schot, J., & Truffer, B. (2002). *Experimenting with sustainable transport: The approach of strategic niche management*. London: Spon Press.

Klerkx, L., Aarts, N., Leeuwis, C. (2009, August). *Dealing with incumbent regimes: Deliberateness and serendipity of agency in rural innovation networks*. Paper presented at ESRS Congress 2009, Vaasa, Finland.

LNV. (2008). *Nota Dierenwelzijn*. The Hague: Ministerie van LNV.

Loorbach, D. (2007). Transition management: New mode of governance for sustainable development. Dissertation. Rotterdam: Erasmus University.

Quist, J. (2007). Backcasting for a sustainable future. The impact after 10 years. Dissertation, Technical University Delft.

Rip, A., & Kemp, R. (1998). Technological change. In S. Rayner & E. L. Malone (Eds.), *Human choice and climate change* (Vol. 2, pp. 327–399). Columbus: Battelle Press.

Roep, D., & Wiskerke, J. S. C. (2006). *Nourishing network. Fourteen lessons learned about creating food supply chains*. Doetinchem: Reed Business Information.

Roep, D., Van der Ploeg, J. D., & Wiskerke, J. S. C. (2003). Managing technical-institutional design processes: Some strategic lessons from environmental cooperatives in the Netherlands. *Netherlands Journal of Agricultural Science, 51*, 95–217.

Rogers, E. M. (1962). *Diffusion of innovations*. Glencoe: Free Press.

Rotmans, J. (2003). *Transitiemanagement, sleutel voor een duurzame samenleving*. Assen: Van Gorcum.

Schot, J., & Geels, F. W. (2008). Strategic niche management and sustainable innovation journeys: Theory, findings, research agenda and policy. *Technology Analysis & Strategic Management., 20*, 537–554.

Smith, A., Stirling, A., & Berkhout, F. (2005). The governance of sustainable socio-technical transitions. *Research Policy, 34*, 1491–1510.
Van der Ploeg, J. D. (2008). *The new peasantries: Struggles for autonomy and sustainability in an Era of empire and globalization*. London: Earthscan.
Weaver, P., Jansen, L., Van Grootveld, G., Van Spiegel, E., & Vergragt, P. (2000). *Sustainable technology development*. Sheffield: Greenleaf Publishing.
Wiskerke, J. S. C., & Van der Ploeg, J. D. (Eds.). (2004). *Seeds of transition. Essays on novelty production, niches, and regimes in agriculture*. Assen: Van Gorcum.

Chapter 8
Factors Affecting the Implementation of Measures for Improving Sustainability on Farms Following the RISE Sustainability Evaluation

Christian Thalmann and Jan Grenz

Abstract The response-inducing sustainability evaluation (RISE) is a method for rapid yet holistic sustainability assessment of agricultural production at farm level. Over 600 farms in 18 countries have been analysed using RISE until 2010.[1] We report on lessons learnt from RISE application, with a focus on practical impact. The analysis as such, despite being comparatively farmer oriented and transparent, at best induces reflection. However, it can serve as a "door opener" to address sustainability issues and worked as an instrument to structure discussion with the farmer. Moreover, if the analysis is an integrated part of a process promoting or developing sustainable yet practicable solutions, the farmers may engage for more sustainable production.

Keywords RISE • Sustainability evaluation • Farm level • Adoption of measures • Participative approach

8.1 Background

Sustainable agriculture (SA) is universally recognised as an indispensable component of sustainable development (UN 1992; for one definition of SA, see FAO 1995). Just like the targeted management of public and private enterprises depends on a reliable accountancy, targeted action towards SA requires relevant and tangible information on whether the farm or sector is heading in the "right" direction. An array of SA

[1] Considering the field experience, the RISE sustainability evaluation (indicator framework, software and farm extension process) has been revised until 2012.

C. Thalmann (✉) • J. Grenz
School of Agricultural, Forest and Food Sciences, Bern University of Applied Sciences, Laenggasse 85, CH-3052 Zollikofen, Switzerland
e-mail: christian.thalmann@bfh.ch; jan.grenz@bfh.ch

A.A. Marta-Costa and E. Silva (eds.), *Methods and Procedures for Building Sustainable Farming Systems: Application in the European Context*,
DOI 10.1007/978-94-007-5003-6_8,

assessment tools have been developed for various scales (Zahm et al. 2008; Meul et al. 2008; Breitschuh et al. 2008; to cite just a few operating at the farm level). The practical implementation of these tools is in an early stage and will likely require further adaptation based on practical experience before unfolding its potential.

8.2 The RISE Approach

RISE is a method for assessing agricultural production at farm level within 1 year (Häni et al. 2008). It starts with the collection of comprehensive information on ecological, economic and social aspects through a questionnaire-based interview with the farmer. A computer model uses this information to calculate 57 sustainability parameters, condensed into 12 indicators (Table 8.1).

All RISE indicators are composed of state (current situation of the system) and driving-force (pressures on the system) parameters. The indicator degrees of sustainability are calculated as the aggregate state minus the aggregate driving force. Indicators are rated on a scale from −100 to +100, where +100 indicates the optimum and −100 an intolerable situation. Benchmark values used for normalisation onto this scale are derived from published research, official statistics (e.g. FAOSTAT), (inter-) national agreements (e.g. ILO decent work standards) and expert knowledge. Some benchmarks can be regionally adapted. An optimum situation in the sense of farm sustainability is reflected by a broad bandwidth of positive indicator values rather than a maximisation of single indicators. Indicator scores are visualised as a polygon (see Figs. 8.2, 8.4, 8.7). At parameter level, results are presented in tabular form to allow for a more detailed presentation. The RISE method has been developed iteratively through a sequence of development, application, evaluation and improvement phases since 2000. The current version 1.1 is being thoroughly revised based on the experiences described in this chapter and in cooperation with extension and communication experts and farmers in order to optimise the indicator set, increase flexibility and user-friendliness of the software and facilitate the integration of RISE into a solution-oriented knowledge management system.

8.3 Previous Applications of RISE

Up to now, RISE has been applied in a large variety of contexts: collaborative projects, education and training modules, research and development studies involving private industry, political and research institutions, producer organisations and farmers. The comprehensiveness of the sustainability evaluation varied depending on the objectives of the respective project: (i) Only the analysis was done, (ii) the analysis was supplemented by a detailed feedback to the farmer and, finally, (iii) processes such as training programmes or the establishment of a platform for sustainable agriculture were initiated (Fig. 8.1).

Table 8.1 Indicators and parameters of the RISE method for evaluating the sustainability of agricultural production

Indicators	State parameters	Driving-force parameters
Energy	Environmental effects of energy carriers used	Energy input per unit agricultural land
		Energy input per unit workforce
Water	Water quantity and stability of the quantity	Water quantity and productivity (crop and animal production)
	Water quality and stability of the quality	Risks to water quality (manure, silage leachate, wastewater, etc.)
Soil	Soil pH, salinisation, waterlogging, soil sampling	Pollution by pesticides, acidifying fertilisers and fertilisers containing heavy metals
		Tillage-related risks
	Erosion index	Salinisation risk
		Nutrient mining
Biodiversity	Biodiversity-promoting practices	Proportion of intensively used agricultural land
		Plot size
		Weed control
N and P emission potential	N and P balance	N and P from organic and inorganic fertilisers (imports/exports)
	Manure storage and application	
Plant protection	Quality of the application	Crop husbandry
	Eco- and human-toxicological risks	Crop rotation
Waste	Environmental hazard	Type and quantity of waste
	Methods of waste disposal	
Economic stability	Net debt service over change in owner's equity and interest paid	Cash flow/raw performance rate
	Equity ratio	Dynamic gearing
	Gross investment	Condition of machines, buildings and perennial crops
Economic efficiency	Return on assets	Productivity
	Return on equity	
	Total earned income	
Local economy	Share of regional workforce and salaries	Raw performance per unit agricultural land
	Lowest salary on farm compared to average regional salary	
Working conditions	Emergency/medical care	Continuing education
	Provision of potable water	Encumbering work
	Accommodation and sanitary equipment	Working conditions
	Working hours	
	Wage discrimination	Income disparity
	Child labour	
	Forced labour	Working time to reach minimum salary
	Gender	
Social security	Social security	Potentially payable salary
		Farm succession plan
	Means of subsistence	Legality and documentation of employment

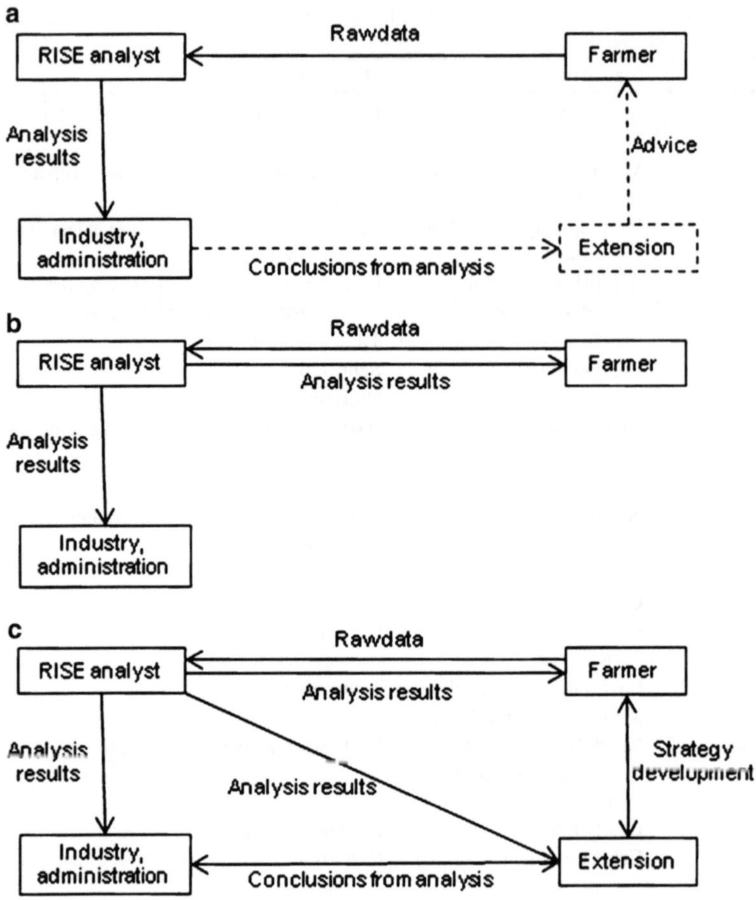

Fig. 8.1 Typical contexts of RISE application: (**a**) Analysis without feedback. (**b**) Analysis supplemented by a detailed feedback to the farmer. (**c**) Analysis, feedback and follow-up process, e.g. training programmes or the establishment of a platform for sustainable agriculture. *Solid lines* represent standard flow of information and actors in projects. *Dashed lines* represent optional flows of information and actors, which only existed in part of the projects

8.4 "Pure Analysis" (One-Way Flow of Information)

In certain projects, the on-farm part of the RISE procedure was restricted to data collection (Fig. 8.1a). During the visit, the analyst conducted a standardised interview with the farmer, following a 20-page questionnaire. The interview topics covered animal husbandry; the use of water, pesticide and fertiliser; plot-level information on topography, soils and biodiversity; the crop calendar; working conditions; social security and farm financials. The order of these topics in the interview followed the logic that less controversial topics like energy use were addressed earlier, sensitive topics towards the end of the questionnaire. This should give time to create an atmosphere of trust before turning to social and financial issues. The analyst had to be an expert in agronomy and command good knowledge of the local environmental, sociocultural, economic and agronomic contexts. Particularly in Switzerland, farmers' willingness to actively contribute to the analysis was strongly tied to the perceived competence of the interviewer. When only a pure analysis was done, the results were solely presented to and discussed with the project sponsors which usually were private companies or administration but not with the farmers themselves. This procedure was chosen when the focal point of the project was not the improvement of an individual farm but rather the quick capturing of the sustainability situation of a group of farms representative of a region, country or catchment area of a processing factory. Following this procedure, an implementation of measures at farm level as a direct consequence of the analysis could hardly be observed. However, project stakeholders tackled identified sustainability deficits by their own possibilities and means. For example, in northern China, the installation of training programmes and strengthening of a regional extension service were implemented by a private company and government, following a RISE study (Box 8.1).

Box 8.1 Example of a "pure analysis" RISE project: Heilongjiang Province, China (2002, 2009)

In a cooperative project of the Swiss College of Agriculture, a private food company, the regional Chinese government and the Northeast Agricultural University of Harbin, RISE was used to assess the sustainability of dairy farms in north-eastern China. The aim was to identify key sustainability issues in the catchment area of a newly established dairy factory. Farmers did not receive a direct feedback of the RISE results, which were instead presented to and discussed with representatives of the government and private company.

Farm characterisation (average values):

Number of farms analysed: 30
Agricultural area: 1.46 ha
Work forces: 1.45
Large livestock units: 4.8
Animals: dairy cows
Main crops: maize (fodder), vegetables
Equity ratio: 0.96
Annual farm income: 2,681 US$
Net profit/loss: 1,377 US$

Major issues at parameter level:

Remark: Indicator values (polygon) are aggregated parameter values (see Fig. 8.2).

- Excessive stocking rates (*biodiversity* indicator; see summary polygon below)
- Use of mineral fertilisers in addition to an existing surplus of manure (*N and P potential* indicator)

(continued)

Box 8.1 (continued)

- Improper manure storage (leaching) (*N and P potential* indicator) (Fig. 8.3)
- Inappropriate skills concerning application of plant protection products (*plant protection* indicator)
- Insufficient social securities (*social situation* indicator)

Outcome and follow-up process:
Following the RISE study, the regional government and the private company fostered activities in the identified priority domains. The development of a consulting team as well as the construction of biogas digesters and the production of corn silage and alfalfa were some of these actions. A team of a private agricultural service in the district provided technical assistance. A training video was produced that informed farmers about proper feeding, hoof treatment, silage production and general hygiene. Brochures with further information for farmers were developed. Additionally, posters with information on feeding and other aspects were hung up in the milk collection centres. In 2009, some of the farmers were reassessed with RISE in order to monitor the development of the farms. Most farmers mentioned not having implemented any changes after the first RISE analysis. However, partly as a consequence of the RISE analysis, several measures had been proposed by the extension services and local government to these farmers.

One important issue was that many interview questions did not allow the farmer to draw direct conclusions about the outcome of the analysis. They rather induced reflection, as confirmed by a survey among Swiss farmers who had previously participated in RISE studies. From the experiences made, we induce that in settings without comprehensive follow-up to the sustainability analysis, (i) all questions asked should be clearly linked to sustainability topics, (ii) the reason for asking these questions should be explained to the farmer, (iii) preliminary results should be shared with the farmer during the first visit and (iv) farmers have to be informed about their role in the project and the objectives of the latter.

8.5 Analysis and Feedback (Two-Way Flow of Information)

The full procedure of a RISE analysis consisted of two visits to the same farm (Fig. 8.1b). The second visit served the presentation of the analysis results to the farmer, the joint control of the plausibility of results and the discussion of possible further steps. RISE results were compiled and presented in a feedback booklet at different levels of detail.

This allowed the communication of results to the farmer to be flexibly adapted as the feedback discussion unfolded. However, according to feedbacks from farmers, there was a rather poor implementation rate of sustainable agricultural practices following the analysis cum feedback procedure, at least within the first years after the analysis (Box 8.2).

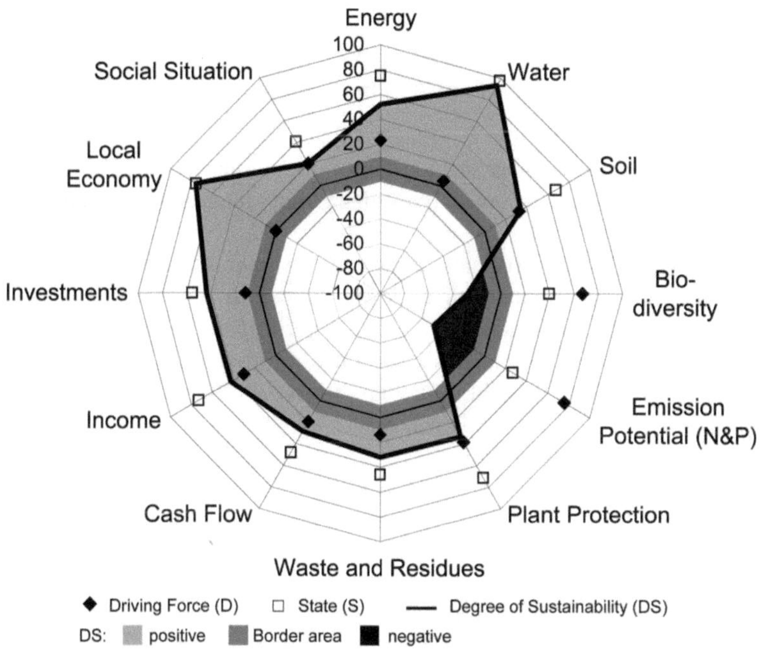

Fig. 8.2 Summary polygon with mean indicator degrees of sustainability of 30 dairy farms in Heilongjiang Province, China

Fig. 8.3 Cattle sheds and inappropriate manure management in Heilongjiang Province, China, 2002

Box 8.2 Example of RISE project (analysis and feedback): in four regions in the Swiss Alps (cantons of Grisons, Bern, St. Gallen, Lucerne) (2006)

In the "Mountain Dairy Project", strategies for commercial dairy manufacturing plants in mountainous areas and for dairy milk producers were developed in order to improve competitiveness of the dairy sector in an expected deregulated cheese market.

RISE was used to holistically assess farm situations and to validate the individual farm development strategies developed by another project group (economists). The RISE analyst visited the farm twice, once for collecting data for the sustainability analysis and once to provide an individual feedback. There was no consistent follow-up process after the feedback.

Farm characterisation (average values):
Number of farms analysed: 10
Agricultural area: 30.3 ha
Work force: 1.8
Number of large livestock units: 27.4
Animals: dairy cows
Main crops: pastures, meadows
Equity ratio: 0.49
Farm income: 66,600 US$
Net profit/loss: −16,246 US$

Major issues at parameter level:
Remark: Indicator values (polygon) are aggregated parameter values (see Fig. 8.4).

- Critical water availability due to dry climatic conditions in some locations (*water* indicator)
- High energy consumption (*energy* indicator)
- High production costs (high labour costs) and low profitability (*economic efficiency* indicator)
- Ammonia evaporation due to problematic manure application practices (*N and P emission potential* indicator) (Fig. 8.5)

Outcome and follow-up process:
The farm strategies developed in this project were mainly oriented towards economic aspects and generally suggested an intensification of production (e.g. by increasing the number of dairy cows). Off-farm income was to be increased. The independent, holistic RISE evaluation provided a complementary picture and triggered discussion about otherwise ignored aspects of sustainability. Examples of issues that were brought up were workload, ecological aspects of irrigation and cross-subsidisation of agricultural production by off-farm activities. Within the frame of this project, it was not possible to develop viable strategy alternatives, since limited resources made the required in-depth analysis impossible. For an optimal consideration of sustainability issues in

(continued)

Box 8.2 (continued)

farm strategy planning, it would be essential to consider a holistic baseline analysis at an earlier stage of the process. Sufficient resources should be reserved for the complex and time-consuming development of alternatives.

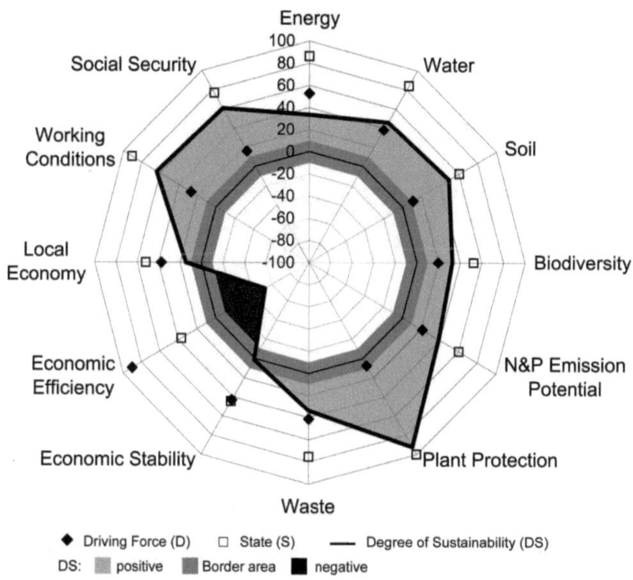

Fig. 8.4 Summary polygon with mean indicator degrees of sustainability of 10 mountain dairy farms in the Swiss Alps

This observation was supported by a small study with Swiss farmers on the implementation of measures following a RISE analysis. Of ten surveyed farmers who had participated in RISE studies in the previous 3 years, eight had not changed anything on their farm. They do "keep the RISE information in mind" and "might take the results into consideration" in specific situations, e.g. when a new farming strategy would be needed due to changes on markets or of the family situation or when a major investment would be considered. Developers and users of indicator-based holistic farm analyses hence should be clear about the importance of adjusting the timing of interventions to farmers' needs.

8.6 Analysis, Feedback and Follow-Up Process (Three-Way Flows of Information)

In projects aiming at the practical implementation of measures fostering agricultural sustainability, the RISE analysis set off a process involving experts with local knowledge (Fig. 8.1c). The RISE analyst completed the analysis, verified the results

Fig. 8.5 Liquid manure application with splash plate on a dairy farm in the Swiss Alps, 2006

with the farmer and provided information about the sustainability situation to an extension service. The latter merged the RISE information with local knowledge and provided specific, solution-oriented advice to farmers. A structure of this type was implemented in a collaborative project in Kenya (see Box 8.3). The overall goal was to contribute to improving the situation of smallholder farmers by making agricultural production more sustainable. RISE was used for a holistic analysis (1) to prioritise issues and identify entry points for concrete measures and (2) to monitor the impact of training measures over time (2009 versus 2006).

Together with local agronomists and institutions, training modules on priority domains such as record keeping, soil conservation and manure management were developed and training courses accomplished, to which 240 farmers attended. Demonstration plots were established and regular follow-up meetings and trainings organised to foster long-term relationships between the regional extension team and the farmers. The RISE results offered a comprehensive basis for a data-based dialogue on sustainability issues. The 2009 re-evaluation of the 30 farms analysed in 2006 revealed good adoption of SA practices with immediate and visible effects that either alleviated workload (use of herbicides instead of hand weeding) or mitigated the risks of crop failure in this drought-prone region (water harvesting, irrigation, no/reduced tillage). Measures with rather diffuse cause-effect relations (e.g. manure management, record keeping) were virtually not adapted.

Box 8.3 Example of RISE project (analysis, feedback and follow-up process): Nanyuki (Laikipia District), Kenya (2006, 2009)

In a collaborative project of HAFL (School of Agricultural, Forest and Food Sciences), CETRAD (Centre for Training and Integrated Research in Arid and Semi-arid Lands Development) and Syngenta, key sustainability deficits of smallholder agriculture in the Laikipia District of Kenya (Fig. 8.6) were identified through RISE assessments and tackled through targeted training modules. The overall goal of the project was to improve the situation of smallholder farmers by enhancing the sustainability of their agricultural production.

Those findings applying to all farms were brought up in group feedback discussions. Farmers were invited to a farmer's site, a community hall or church, and results of general interest, like crop rotation or water use, were presented. Individual feedback on specific and more sensitive issues, like profitability or social security, was also provided to all interviewed farmers.

Farm characterisation (average values):
Number of farms analysed: 30
Agricultural area: 3.21 ha
Work forces: 2.8
Number of large animal units: 3.98
Type of animals: (dairy) cattle, goats, poultry
Crops: maize, potatoes, beans, wheat
Equity ratio: 0.98
Farm income: 1,604 US$
Net profit/loss: −3,402 US$

Major issues at parameter level:
Remark: Indicator values (polygon) are aggregated parameter values (Fig. 8.7).

- Lack of water due to extreme climatic conditions and overuse of water resources (*water* indicator)
- Poverty and even malnutrition prevail due to low crop yields and crop failure (*water* indicator)
- Production techniques that enhance soil water evaporation (*soil* indicator)
- Inappropriate manure management (*N and P emission potential* indicator)
- Inappropriate skills concerning application of plant protection products (*plant protection* indicator)
- Insufficient social securities (*social security* indicator)

Outcome and follow-up process:
In a follow-up process, an extension service team was established, and trainings on selected topics were offered. Farmers throughout appreciated the holistic "portfolio" of their operation presented in the feedback. In contrast to precedent trainings that focused on the most common issues, the feedbacks allowed for thematically open discussions. Advantages of working in groups were the

(continued)

Box 8.3 (continued)

lively discussions and the exchange of experience and knowledge that promoted collaborations among the participants.

In 2009, the farms were re-evaluated by the same RISE analyst. Several farmers had adopted conservation agricultural (CA) techniques on their farm or participated in trainings for safe use of chemicals. Repeated crop failure due to drought motivated several farmers to try new techniques like no-till, reduced tillage or water harvesting. Demonstration plots and farms and repeated visits of extension officers supported adoption. However, for many farmers, it was difficult to transfer innovations presented, e.g. on field days or demo farms onto their own farm. The challenge is to join the new elements with the existing farm strategy. For example, farmers who want to produce crops with CA techniques should cover the soil with mulch to prevent unproductive evaporation. Some farmers did use mulch on their plots but did not adjust the number of livestock to the reduced amount of biomass available. The consequence was that not enough mulch was available for an effective soil cover and the fodder for livestock became critical. Optimally, the extension service individually assists implementation of measures at each farm.

Fig. 8.6 Smallholder farms with small plots in Laikipia District, Kenya, 2009

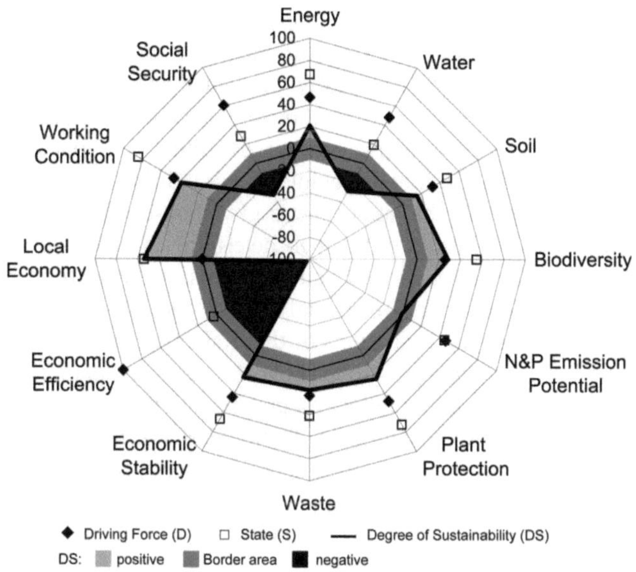

Fig. 8.7 Summary polygon with mean indicator degrees of sustainability of 30 smallholder farms in the Laikipia District, Kenya

8.7 Key Factors for an Efficient Implementation of Measures

The implementation of measures varied considerably between projects depending on context and procedure of the RISE application:

- Generally, farmers only got involved in SA if the sustainability analysis was an integrated part of a process generating sustainable yet practicable solutions. Some of the re-evaluated Swiss farmers stated that the RISE analysis alone was biased, e.g. due to the system boundary's excluding off-farm income and some of the economic parameters being based on tax accounting (which usually is "tax optimised" and hence based on a too low revenue). The trade-off between fast and broad applicability of the tool and specific relevance of the results affects all known comparable indicator systems (von Wirén-Lehr 2001). Neither the sustainability parameters nor the benchmark values used in RISE will equally satisfy the demands of all the various actors with a stake in SA (Pretty et al. 2008). As a consequence, the next RISE version will keep a universal set of indicators and parameters but allow for a higher degree of benchmark adaptation according to user requirements. Particularly in the socio-economic domain, farmers' perceptions, e.g. on the relative importance of farm and off-farm income, should be reflected in the weighting of parameters. Implementation of this participatory approach to benchmark selection should alleviate the outlined dilemma of applicability and specificity (von Wirén-Lehr 2001; Grunwald and Kopfmüller 2006).

- Farmers should have the chance to develop genuine interest for a sustainable development in general and SA in particular. A transparent definition of SA is necessary, translated into practically implementable goals (von Wirén-Lehr 2001). Valuation algorithms must be directly related to these goals. Since sustainable development is a normative concept, definition and goal system of SA are necessarily subjective to a considerable extent (Christen 1999). Environmental limits, such as the tolerable atmospheric greenhouse gas concentration, can be considered absolute, albeit difficult to determine. Yet, in the socio-economic domain, subjective concepts and value systems stand in the centre of human behaviour. It is important to transparently present the values on which an indicator system is based and the limitations of the evaluation as well as to periodically review these values together with the concerned stakeholders.
- Farmer's involvement in the follow-up process and the search for practical solutions should be actively encouraged. Techniques, e.g. from the empirical social sciences, such as participatory rural appraisal, may prove useful in this regard (e.g. Chambers 1994). Combined approaches involving "hard science" and "soft communication" seem favourable: scientific knowledge adapted to the specific situation (Stähli and Egli-Schaft 2008).
- Projects involving sustainability analysis resp. targeting SA implementation should be more systematically evaluated with respect to their defined goals. These goals should be formulated together with the concerned stakeholders and considering established standards and methods (e.g. IFAD 2009).

8.8 Establishment of a Knowledge Platform

The availability of user- and solution-oriented information proved an important factor in the adoption of SA techniques (see also Rodriguez et al. 2009). A central challenge in making sustainability-relevant knowledge effective is to bring the rather abstract concept of sustainable development to the ground. For this purpose, the RISE team is working on a project aiming at developing a knowledge management system that identifies sustainability deficits and connects people, knowledge and technologies for practicable SA solutions. This knowledge management system shall neither be a decision support system nor an expert model, given the often limited consideration of field reality and farmer's knowledge and experience by such systems. It rather will be a knowledge and communication platform of a knowledge community linking farmers and other stakeholders.

References

Breitschuh, G., Eckert, H., Matthes, I., & Strümpfel, J. (2008). *Kriteriensystem nachhaltige Landwirtschaft (KSNL) (in German)*. Darmstadt: Kuratorium für Technik und Bauwesen in der Landwirtschaft.

Chambers, R. (1994). The origins and practice of participatory rural appraisal. *World Development, 22*, 953–969.

Christen, O. (1999). *Nachhaltige Landwirtschaft – Von der Ideengeschichte zur praktischen Umsetzung (in German)*. Bonn: FIL.

Food and Agriculture Organisation of the United Nations (FAO). (1995). *Trainer's manual* (Sustainability issues in agricultural and rural development policies, Vol. 1). Rome: Food and Agriculture Organisation.

Grunwald, A., & Kopfmüller, J. (2006). *Nachhaltigkeit (in German)*. Frankfurt/New York: Campus-Verlag.

Häni, F., Studer, C., Thalmann, C., Porsche, H., Stämpfli, A. (2008). RISE – Massnahmenorientierte Nachhaltigkeitsanalyse landwirtschaftlicher Betriebe (in German). KTBL-Schrift 467. Darmstadt: Kuratorium für Technik und Bauwesen in der Landwirtschaft.

International Fund for Agricultural Development (IFAD). (2009). *Evaluation manual, methodology and procedures*. Rome: Office of Evaluation, IFAD.

Meul, M., Van Passel, S., Nevens, F., Dessein, J., Rogge, E., Mulier, A., & Van Hauwermeiren, A. (2008). MOTIFS: A monitoring tool for integrated farm sustainability. *Agronomy of Sustainable Development, 28*, 321–332.

Pretty, J., Smith, G., Goulding, K. W. T., Groves, S. J., Henderson, I., Hine, R. E., King, V., van Oostrum, J., Pendlington, D. J., Vis, J. K., & Walter, C. (2008). Multi-year assessment of Unilever's progress towards agricultural sustainability I: Indicators, methodology and pilot farm results. *International Journal of Agricultural Sustainability, 6*, 37–62.

Rodriguez, J. M., Molnar, J. J., Fazio, R. A., Sydnor, E., & Lowe, M. J. (2009). Barriers to adoption of sustainable agricultural practices: Change agent perspectives. *Renewable Agriculture and Food System, 24*, 60–71.

Stähli, R., & Egli-Schaft, W. (2008). Aktionsforschung – eine Forschungsmethode auch für die Landwirtschaft (in German). *Agrarforschung, 15*, 4–6.

United Nations (UN). (1992). Agenda 21. Environment and Development Agenda. www.unep.org/Documents.Multilingual/Default.asp?documentID=52. Accessed 10 January 2012.

von Wirén-Lehr, S. (2001). Sustainability in agriculture – An evaluation of principal goal-oriented concepts to close the gap between theory and practice. *Agriculture, Ecosystem & Environment, 84*, 115–129.

Zahm, F., Viaux, P., Vilain, L., Girardin, P., & Mouchet, C. (2008). Assessing farm sustainability with the IDEA method – From the concept of agriculture sustainability to case studies on farms. *Sustain Development, 16*, 271–281.

Chapter 9
A Multi-attribute Decision Method for Assessing the Overall Sustainability of Crop Protection Strategies: A Case Study Based on Apple Production in Europe

Patrik Mouron, Ursula Aubert, Bart Heijne, Andreas Naef, Jörn Strassemeyer, Frank Hayer, Gérard Gaillard, Gabriele Mack, José Hernandez, Jesus Avilla, Joan Solé, Benoit Sauphanor, Aude Alaphilippe, Andrea Patocchi, Jörg Samietz, Heinrich Höhn, Esther Bravin, Claire Lavigne, Marko Bohanec, and Franz Bigler

Abstract In this study, we investigated the elements that must be considered to obtain a clear and useful assessment of sustainability. We present a system-description tool created especially for life cycle assessment (assessment of energy use and ecotoxicity), environmental risk assessment, and full-cost calculations. Using the various results from these assessments as qualitative attributes, we designed a multi-attribute tool that allows us to integrate sustainability attributes over five levels into

P. Mouron (✉) • U. Aubert • F. Hayer • G. Gaillard • G. Mack • J. Hernandez • F. Bigler
Research Station, Agroscope Reckenholz-Tänikon,
CH-8046 Zürich, Switzerland
e-mail: patrik.mouron@art.admin.ch

B. Heijne
Applied Plant Research, Wageningen UR,
P.O. Box 200, 6670 AE Zetten, The Netherlands
e-mail: Bart.Heijne@wur.nl

A. Naef • A. Patocchi • J. Samietz • H. Höhn • E. Bravin
Research Station, Agroscope Changins-Wädenswil,
CH-8820 Wädenswil, Switzerland
e-mail: andreas.naef@acw.admin.ch

J. Strassemeyer
JKI; Julius Kühn-Institute, Federal Research Centre for Cultivated Plants,
Stahnsdorfer Damm 81, 14532 Kleinmachnow, Germany
e-mail: joern.strassemeyer@jki.bund.de

J. Avilla • J. Solé
Centre UdL-IRTA for R+D, AGROTECNIO, University of Lleida,
Rovira Roure – 191, 25198 Lleida, Spain
e-mail: avilla@pvcf.udl.cat

A.A. Marta-Costa and E. Silva (eds.), *Methods and Procedures for Building Sustainable Farming Systems: Application in the European Context*,
DOI 10.1007/978-94-007-5003-6_9,

an overall sustainability rating. To demonstrate the transparency of this method and how it enables decision makers to deal with complexity, we use the method to assess different crop protection systems used in apple production. Although the multi-attribute decision method provided a reasonable overall assessment of the sustainability of different protection systems, the assessment could be substantially influenced by the selection of rating scales and decision rules. Therefore, the rating scales and decision rules should be carefully defined and discussed among the research teams. In our case, experts have participated from five European countries.

Keywords Multi-attributive decision making • Apple orchard • Crop protection strategy • Sustainable development • Life cycle assessment (LCA) • SYNOPS • Full-cost calculation

9.1 Introduction

European agricultural policy requires the implementation of integrated pest management (IPM) by 2014. The goal is to promote crop protection strategies that are less reliant on pesticides (ENDURE 2009). All members of the EU will have to propose a national action plan to implement IPM strategies adapted to regional conditions. Although methods and tools to evaluate the overall sustainability (including environmental and socio-economic aspects) of such region-based IPM strategies are needed, they are largely unavailable. In contrast, assessments of single aspects of sustainable development have often been published. For environmental aspects of the sustainability of agricultural systems, Foster et al. (2006) provide a review for European countries, mainly based on "life cycle assessment" methodology. Methods that include both environmental and socio-economic aspects are provided by the "response-induced sustainability evaluation" or RISE (Grenz et al. 2009) and by the concept of "sustainability solution spaces" (Wiek and Binder 2005; Castoldi et al. 2007). These tools, however, do not attempt to aggregate the various aspects of sustainability into a single rating of the overall sustainability of a system.

Multi-attributive decision making offers a methodological framework for defining hierarchical trees of attributes that generate an overall sustainability rating (Bockstaller et al. 2008; Sadok et al. 2009). This has been demonstrated by Bohanec et al. (2008), who applied a multi-attribute model for economic and ecological

B. Sauphanor • A. Alaphilippe • C. Lavigne
INRA, UERI Domaine de Gotheron,
F-26320 Saint Marcel-lès-Valence, France
e-mail: benoit.sauphanor@avignon.inra.fr

M. Bohanec
Department of Knowledge Technologies, Jožef Stefan Institute,
Jamova cesta 39, SI-1000 Ljubljana, Slovenia
e-mail: marko.bohanec@ijs.si

assessment of genetically modified crops, and by Lô-Pelzer et al. (2009), who evaluated innovative crop protection strategies for arable production systems. These multi-attributive studies share an important characteristic, which is that they facilitate consideration of system complexity. The number of attributes used in these models is very high, i.e. the models often include more than 80 attributes on more than seven hierarchical levels.

Although such large "attribute trees" can easily be handled by computer programs (Bohanec 2011), much effort is required to understand and communicate the cause-and-effect relations contained in such models. Transparency should be enhanced so that the logic in the model can be understood, evaluated, and modified as needed.

The goal of this chapter is to investigate the methodological elements that need to be considered to obtain transparency in sustainability assessment. An example concerning crop protection systems in apple production is used to demonstrate the transparency of this method. The rating scales and decision rules used in the sustainability assessment described here were defined by a group of experts who had participated in the EU-FP6 project ENDURE.

9.2 Scheme for Sustainability Evaluation

We propose a scheme for sustainability assessment of orchard and other cropping systems that includes five elements. The assessment begins by describing the farming-system parameters (Fig. 9.1a). The settings of these parameters are then used to conduct quantitative assessments referring to the main dimensions of sustainability, which are in our case ecology and economics (Fig. 9.1b). The diverse output variables of the assessments or "basic attributes" are then entered at the bottom of a hierarchical attribute tree (Fig. 9.1c). Here, the quantitative results are transformed into qualitative ratings in order to aggregate them into attributes of higher levels (Fig. 9.1d). For optimising crop protection systems, however, we need to know which parameters substantially influence the assessment of overall sustainability (Fig. 9.1e). Such cause-and-effect relations can be obtained by investigating the results from top to bottom in the scheme described in Figure 9.1. In the following sections, we describe the components of Fig. 9.1 in greater detail.

9.3 System-Description Tool

Our study compares crop protection strategies that reduce pesticide application with a baseline system (BS) that strictly relies on pesticide application. We distinguished therefore an advanced system (AS) that replaces pesticides as much as possible by alternative methods that are available on the market and an innovative system (IS) that replaces pesticides by alternative methods that are currently used in field trials or laboratories but are not available on the market. The system descriptions include

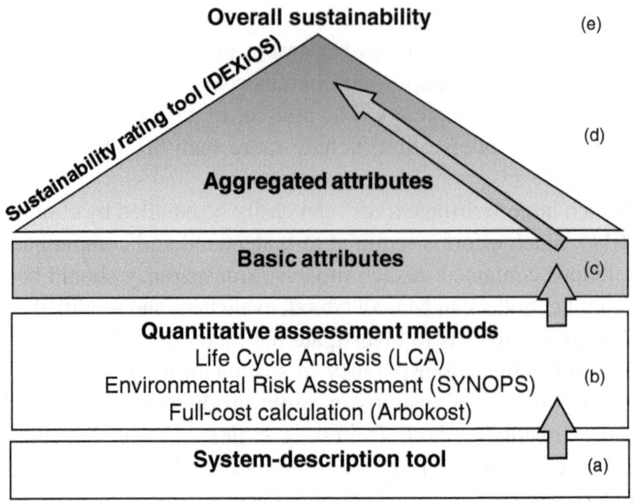

Fig. 9.1 Scheme for assessing the overall sustainability of crop systems

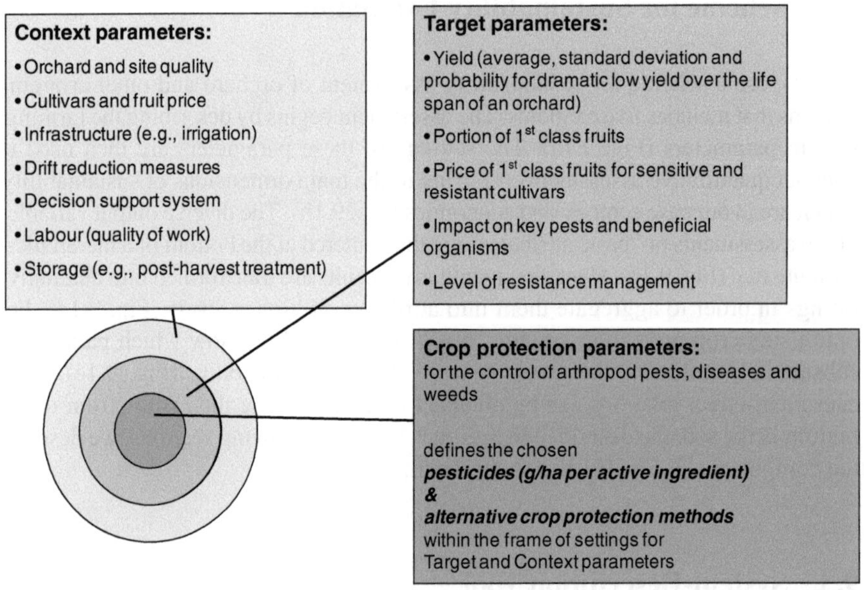

Fig. 9.2 Three types of system-description parameters for defining crop protection strategies for apple production

detailed information concerning the active ingredients applied, the dosages applied, and the time of application (the calendar week). Such parameters must be related to expected yield levels. Expected yields can be estimated with the "target yield approach" (Bera et al. 2006). The target yield approach takes into consideration the efficiency of crop protection parameters for achieving the desired target parameter level (e.g. yield) for a particular orchard system with given context parameters. Figure 9.2 illustrates how the definitions of crop protection parameters are embedded within context parameters and target parameters in our "system-description tool". By keeping context parameters and target parameters for a region constant, we were able to compare the sustainability of different crop protection strategies (i.e. BS vs. AS and IS) while assessing the whole farming system.

9.4 Quantitative Assessment Methods

9.4.1 Life Cycle Assessment (LCA)

The LCA considers not only impacts related to the use of pesticides but also environmental impacts related to pesticide production and transport. The LCA also includes other activities and their related inputs (resource use) in an apple orchard over a season, i.e. fertiliser, machinery, buildings, hail net, and field operations such as harvesting and mulching. LCA does not include the creation and uprooting of the orchard, irrigation, and post-harvest processes like storage.

The design of the LCA follows the principles outlined by ISO (2006). Values from system-description parameters (Fig. 9.2) for apple orchards are transformed into the life cycle inventory, which is used to evaluate the environmental effects. We used the life cycle inventories from the ECOINVENT database version 2.01 (Frischknecht et al. 2007; Nemecek and Kägi 2007; Nemecek and Erzinger 2005) to assess the infrastructure, inputs, and processes used in the apple orchards. The models used to estimate the various direct field emissions (i.e. NH_3, N_2O, P_2O_5, NO_3^-, heavy metals, and pesticides) are described in the SALCA method (Gaillard and Nemecek 2009; Nemecek et al. 2005, 2008).

The following sustainability attributes were derived as part of the LCA in this study: terrestrial and aquatic ecotoxicity potential of toxic pollutants were calculated according to Guinée et al. (2001); human toxicity potential of toxic pollutants via exposure through food, tap water, and air were calculated according to Guinée et al. (2001); demand for non-renewable energy resources was estimated according to Hischier et al. (2009); global warming potential over 100 years was considered as described in IPCC (2006); and eutrophication potential (the impact of the losses of N and P to aquatic and terrestrial ecosystems) was calculated according to the EDIP97 method (Hauschild and Wenzel 1998).

9.4.2 SYNOPS

The indicator model SYNOPS assesses the risk caused by pesticide drift. In particular, the model assesses the risk for organisms living in terrestrial (i.e. soil and field margins) and aquatic (i.e. surface water) habitats. It combines pesticide-use data, including different degrees of drift-reduction measures, with environmental conditions (e.g. distance from the orchard to surface water). Chemical, physical, and ecotoxicological properties of applied active ingredients are taken into account (Gutsche and Strassemeyer 2007). In general, the acute and chronic risk potentials are calculated as exposure-toxicity ratios (ETR) for reference organisms such as earthworms for soil, bees for the above-ground area in the crop and in the crop borders, and daphnia, algae, and fish for surface water. Time-dependent pesticide concentration curves are used to derive the acute and chronic risk potentials by relating pesticide concentration in the environment to the lethal concentration (LC50) and the no-effect concentration (NOEC). For each crop protection system under study, the indicator model SYNOPS was applied to assess the region-specific environmental risk potentials. The region-specific and field-related environmental conditions like slope, soil type, and climate were derived from a spatial database, which was developed within the EU project HAIR (2007).

The following sustainability attributes were derived from the SYNOPS assessment in this study: terrestrial acute risk, terrestrial chronic risk, aquatic acute risk, and aquatic chronic risk.

9.4.3 Full-Cost Calculation

Orchards are capital- and labour-intensive perennial systems. Income may vary considerably among years depending mainly on variation in fruit yield and the proportion of 1st-class fruit (Mouron and Scholz 2007). In addition to calculate average annual income, our economic assessment therefore determines variability in income based on the standard deviation of yield and the proportion of 1st-class fruit as defined in system-description parameters. Dramatic yield loss related to the proportion of years with less than half of the average harvest is also taken into account.

Full-cost principles designed especially for perennial tree crops are applied as described by Mouron et al. (2006b). These full-cost principles evaluate the grower's capacity to amortise or reinvest, and they therefore refer to long-term viability. In particular, cost of production includes all inputs as well as labour costs (those of the grower and of the hired workforce) and depreciation for investments (mainly the cost for establishing the orchard). Total revenue considers only the amount of apples sold and price; the same prices per kilogramme and per fruit class are used for all orchard systems within a region (i.e. premium prices are not considered). Direct payments (i.e. money from the government to promote IPM) are not included in the revenue calculation. These limitations for calculating the revenue were necessary because premium prices and direct payments related to IPM have yet to be realised in most countries.

The following sustainability attributes were derived from full-cost assessment in this study: family income per hour, total production cost per kilogramme of 1st-class apples, net profit per hectare, income variability, invested capital per hectare, and return on investment (i.e. net profit per invested capital).

The calculations were conducted with the managerial-economic software tool Arbokost (Arbokost 2009). This full-cost calculation tool is designed especially for perennial crops.

9.5 Sustainability-Rating Tool

9.5.1 Building a Hierarchical Attribute Tree

The attribute tree was built both from the top-down and from the bottom-up (Fig. 9.3). From the top-down and according to the "areas of protection" described by Udo de Haes and Lindeijer (2002), the direct sub-attributes of *ecological sustainability* are *resource use*, *environmental quality*, and *human toxicity*. With regard to apple production, environmental attributes were chosen according to Mouron et al. (2006a, b) and Mila i Canals et al. (2007).

According to Lô-Pelzer et al. (2009), the sub-attributes of *economic sustainability* are *profitability*, *production risk*, and *financial autonomy*.

From the bottom-up, the basic ecological attributes were derived from the LCA and SYNOPS. Because the rating of ecotoxicity is the main attribute that is optimised in this research, the ecotoxicity attribute has many sub-attributes. The basic attributes concerning the economic sustainability of orchard systems were selected based on previous studies (Mouron and Scholz 2007; Bravin and Kilchenmann 2010).

9.5.2 Rating Basic Attributes

The numeric values derived from the assessment methods must be rated as to whether they differ substantially from a baseline system (BS). We used five classes for rating basic and aggregated attributes: much worse than BS, worse than BS, similar to BS, better than BS, and much better than BS.

Basic attributes with strictly positive numeric values require a rating scale that prevents the change of the rating with a shift in the reference system (i.e. a shift in BS). Therefore, the boundary between similar and better is the reciprocal of that between similar and worse, and the boundary between better and much better is the reciprocal of that between worse and much worse. Figure 9.4 shows the asymmetric rating scales we used for LCA results according to Nemecek et al. (2005).

The range for the class "similar" is wider for ecotoxicity and human toxicity attributes than for nutrient and resource management attributes because the methodologies for assessing ecotoxicity are less reliable than those for assessing nutrition and resource management.

For basic attributes that can potentially have negative or positive numeric values (*family income*, *net profit*, and *return on investment*), we used symmetric rating

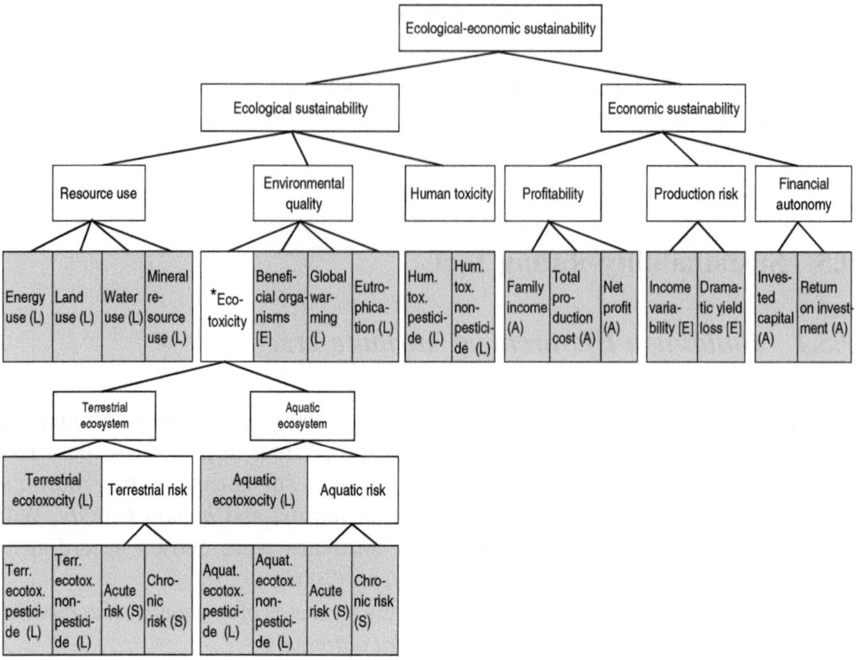

Fig. 9.3 Hierarchical attribute tree for assessing the ecological and economic sustainability of orchard systems. Basic attributes are in *grey boxes*. *Ecotoxicity is the main attribute that is optimised in this research. Letters in parentheses refer to the assessment method: *L* life cycle assessment, *S* SYNOPS, *A* Arbokost, *E* expert estimation

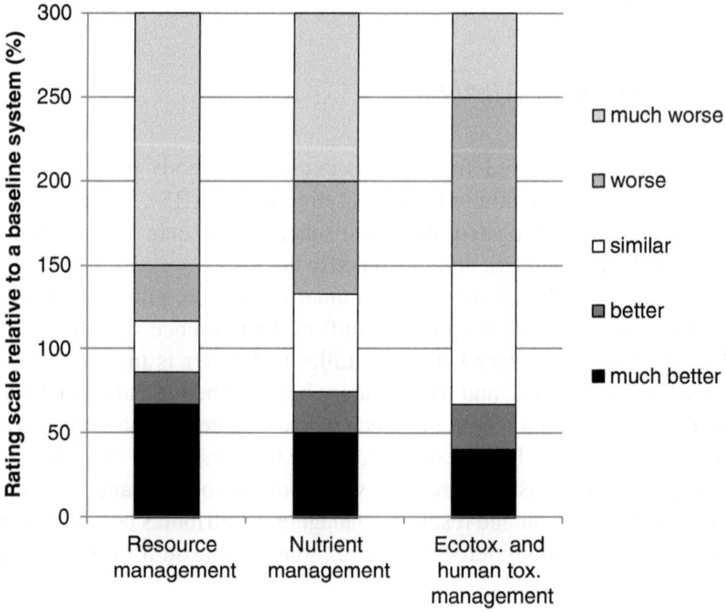

Fig. 9.4 Asymmetric scales for rating life cycle assessment results in relation to a baseline system (=100%)

scales, assuming that a deviation from the reference system (i.e. BS = 100%) in the desired direction is of the same magnitude as a similar deviation in the undesired direction. Here is an example of a symmetric rating scale: similar to BS = 90–110%, better than BS = 110–140%, and worse than BS = 60–90%.

9.5.3 *Rating Aggregated Attributes*

In multi-attribute models, decision rules define how the many sub-attributes are aggregated into one assessment of an attribute (Bohanec et al. 2008). Each aggregate attribute in the model (Fig. 9.3) has an associated set of rules that determine how the aggregation is done. In principle, the rules represent attitudes and preferences of the decision makers; in our case, the rules were specified jointly by experts from five European countries, who were partners in the EU-FP6 project ENDURE.

Table 9.1 shows an example of decision rules that aggregate two sub-attributes into an aggregate attribute. In this case, the two sub-attributes contribute equally to the aggregate attribute; consequently, they are of equal importance and have equal weights. Further, it is assumed that if the two sub-attributes do not differ in their classes for a particular rule (e.g. if both are rated as "similar" to BS), the aggregated attribute will have the same rating class as its sub-attributes (Table 9.1, Nos. 1, 7, 13, 19, 25). If the ratings for two sub-attributes differ by two to four classes, the aggregated attribute will be assigned the class between those of the sub-attributes (Table 9.1, Nos. 3, 5, 9, 11, 12, 15, 17, 21, 23). In all other cases, the assumed rule for aggregation is shown in Table 9.1 (Nos. 2, 4, 6, 8, 10, 14, 16, 18, 20, 22, 24).

9.6 Example of an Overall Sustainability Rating

We compared different apple protection systems under European conditions with the goals of reducing ecotoxicity and maximising overall sustainability. Therefore, we defined a baseline system (BS), an advanced system (AS), and an innovative system (IS). The BS operates only with pesticides within the framework of good agricultural practice. The AS aims to replace pesticides as much as possible with available alternative methods, and the IS has the same goal but also uses alternative methods that are currently used in field trials but that will not be on the market for 10–20 years. Both AS and IS represent integrated pest management principles (IPM). The following assumptions for the crop protection parameters were made:

- Arthropod control:
  - Alternative methods applied for AS and IS: mating disruption, attract and kill, microbial control, sanitary methods, mass trapping, exclosure netting, and predators and parasitoids

Table 9.1 Decision rules for rating aggregated attributes with equal weights

Decision rule number	Sub-attribute 1 e.g., Aquatic ecotoxicity related to pesticide inputs	Sub-attribute 2 e.g., Aquatic ecotoxicity related to non-pesticide inputs	Aggregated attribute e.g., Aquatic ecotoxicity (related to pesticide and non-pesticide inputs)
1	Much worse	Much worse	Much worse
2	Much worse	Worse	Much worse
3	Much worse	Similar	Worse
4	Much worse	Better	Similar
5	Much worse	Much better	Similar
6	Worse	Much worse	Much worse
7	Worse	Worse	Worse
8	Worse	Similar	Similar
9	Worse	Better	Similar
10	Worse	Much better	Similar
11	Similar	Much worse	Worse
12	Similar	Worse	Similar
13	Similar	Similar	Similar
14	Similar	Better	Similar
15	Similar	Much better	Better
16	Better	Much worse	Similar
17	Better	Worse	Similar
18	Better	Similar	Similar
19	Better	Better	Better
20	Better	Much better	Much better
21	Much better	Much worse	Similar
22	Much better	Worse	Similar
23	Much better	Similar	Better
24	Much better	Better	Much better
25	Much better	Much better	Much better

Five rating classes were applied for the two sub-attributes and the aggregated attribute (much worse, worse, similar, better, much better, in relation to a baseline system). If equal weights are not used for the sub-attributes, the decision rules will differ from the example in this table

  - Number of insecticide applications: BS = 12, AS = 8, and IS = 4
- Disease control:
  - Alternative methods applied for AS and IS: resistant cultivars, sanitation, and antagonistic microorganisms
  - Number of fungicide applications: BS = 7, AS = 4, and IS = 3
- Weed control:
  - Alternative methods applied for AS and IS: cover crop from mid-June to harvest with mowing and mechanical weeding
  - Number of herbicide applications: BS = 3, AS = 2, and IS = 2

The sustainability assessment was conducted with the program DEXi (Bohanec 2011). We used the previously described hierarchical attribute tree (Fig. 9.3), rating scales (Fig. 9.4), and decision rules (example in Table 9.1). The resulting ratings for the sustainability attributes are presented in Table 9.2. The ratings indicate that in this example, the *ecological-economic overall sustainability* (attribute No. 1) did not differ substantially between AS, IS, and BS, i.e. both AS and IS were "similar" to BS. This might seem surprising because AS and IS considerably reduced the applications of pesticides compared to BS. We can now investigate the reasons for this outcome. First, the rating of the attribute *ecotoxicity* (Table 9.2, No. 9) was improved by AS and IS as expected; the rating was improved by one class with AS and by two classes with IS. This is mainly due to improvements among the sub-attributes of *ecotoxicity* (i.e. attribute Nos. 10–23). However, *environmental quality* (Table 9.2, No. 8), which is one level higher in the attribute tree, did not differ for AS and BS. This lack of difference is explained by the ratings of the three sub-attributes of *environmental quality*, namely, *impact on beneficial organisms*, *global warming potential*, and *global eutrophication* (Table 9.2, Nos. 24–26). *Environmental quality* contributes together with *resource use* and *human toxicity* to the top attribute of the environmental branch of the tree, which is *ecological sustainability* (Table 9.2, No. 2). On this level, AS remains similar to BS, and IS is rated higher by one class. When the rating for *ecological sustainability* is considered together with the rating from the top attribute of the economic branch, i.e. *economic sustainability* (Table 9.2, No. 30), it is clear that the AS got a rating of "similar" for the overall sustainability because both sub-attributes of overall sustainability were rated "similar". In the case of IS, one sub-attribute of overall sustainability was rated with "similar" and the other was rated "better". According to the decision rules of Table 9.1, the aggregated rating will then be "similar". We point out that the decision rules of Table 9.1 were those that we selected for this example. It would also be possible to define the decision rule as "similar and better=better". As a consequence, the rating of the overall sustainability of IS would be rated higher for one class. This demonstrates the importance of the choice of decision rules in generating aggregate ratings.

9.7 Conclusions

Using apple production in Europe as an example, we have shown how complex systems that include many attributes can be assessed for overall ecological and economic sustainability. We emphasise that the result of such a multi-attribute sustainability assessment might be substantially different depending on definitions and settings of several elements. To obtain transparency of the assessment results, we identified the following tasks:

1. A well-structured system-description tool must be developed to define and control the size of the attribute tree. Defining crop protection parameters in relation to fixed context and target parameters helps decision makers interpret the outcome of the assessment.
2. Established assessment methods such as life cycle assessment, SYNOPS, and full-cost calculation should be applied to ensure that the quantitative analysis is state of the art. Use of these methods also ensures that the models underlying these calculations and the associated uncertainties are clearly described.

Table 9.2 Example for sustainability rating of three apple protection systems

No.	Attribute	Advanced system (AS)	Innovative system (IS)
1	Ecological-economic overall sustainability	Similar	Similar
2	Ecological sustainability	Similar	*Better*
3	Resource use	Similar	Similar
4	Energy use per ha (LCA)	Similar	Similar
5	Land use (LCA)	Similar	Similar
6	Water use per ha (LCA)	Similar	Similar
7	Mineral resource use per ha (LCA)	Similar	Similar
8	Environmental quality	Similar	*Better*
9	Ecotoxicity	Better	*Much better*
10	Terrestrial ecosystem quality	Better	*Much better*
11	Terrestrial ecotoxicity potential (LCA)	Much better	Much better
12	Terrestrial ecotoxicity pesticide (LCA)	Much better	Much better
13	Terr. ecotoxicity non-pesticide (LCA)	Much better	*Better*
14	Terrestrial risk (SYNOPS)	Similar	*Much better*
15	Acute terrestrial risk (SYNOPS)	Similar	*Much better*
16	Chronic terrestrial risk (SYNOPS)	Similar	*Better*
17	Aquatic ecosystem quality	Better	*Much better*
18	Aquatic ecotoxicity potential (LCA)	Better	*Much better*
19	Aquat. ecotox. pot. pesticide (LCA)	Much better	Much better
20	Aquat. ecotox. pot. non-pesticide (LCA)	Similar	*Much better*
21	Aquatic risk (SYNOPS)	Better	*Much better*
22	Acute aquatic risk (SYNOPS)	Better	*Much better*
23	Chronic aquatic risk (SYNOPS)	Better	*Much better*
24	Impact on beneficial organisms	Similar	*Better*
25	Global warming potential (LCA)	Similar	Similar
26	Global eutrophication potential (LCA)	Similar	Similar
27	Human toxicity (LCA)	Better	Better
28	Human toxicity pesticide (LCA)	Much better	Much better
29	Human toxicity non-pesticide (LCA)	Similar	Similar
30	Economic sustainability	Similar	Similar
31	Profitability	Worse	*Similar*
32	Family income per labour hour	Worse	*Better*
33	Total production cost per kg 1st-class fruit	Similar	Similar
34	Net profit per ha	Worse	*Similar*
35	Production risk	Similar	*Better*
36	Income variability	Worse	*Similar*
37	Probability of dramatic yield loss	Similar	*Much better*
38	Financial autonomy	Similar	Similar
39	Invested capital per ha	Similar	*Worse*
40	Return on investment per ha	Worse	*Similar*

Differences in the rating classes between AS and IS are in *bold print*. The following five rating classes were used to compare AS and IS with a baseline system (BS): much worse/worse/similar/better/much better. The sub-attributes were assumed to have equal weight

3. For the translation of quantitative assessment results into qualitative rating classes, asymmetric scales need to be defined if the numeric result cannot be less than zero. Developers and user of this approach to sustainability assessment must recognise that the definition of rating scales might substantially influence the overall sustainability rating.
4. The rating of aggregated attributes depends on decision rules because certain combinations of sub-attribute ratings might be interpreted differently according to subjective preferences. Thus, like the definition of rating scales, the definition of decision rules can substantially influence the overall sustainability rating.

We suggest that these four tasks should be defined by research teams. In this study, the knowledge of experts from five European countries was combined.

Acknowledgements This project was partly founded by EU grant (Project number: 031499) under the Sixth Framework Programme and is catalogued as ENDURE. ENDURE is the European Network for the Durable Exploitation of Crop Protection Strategies.

References

Arbokost. (2009). Betriebswirtschaftliches Simulationsprogramm für Obstproduzenten (Full-cost software tool for fruit growers). Agroscope Changins Wädenswil. http://www.arbokost.info-acw.ch. Accessed 4 January 2012.

Bera, R., Seal, A., Bhattacharyya, P., Das, T. H., Sarkar, D., Kangjook, K. (2006). Target yield concept and a framework of fertilizer recommendation in irrigated rice domains of subtropical India. *Journal of Zhejiang University SCIENCE B, 7*(12), 963–968. http://www.zju.edu.cn/jzus/article.php?doi=10.1631/jzus.2006.B0963. Accessed 4 January 2012.

Bockstaller, C., Guichard, L., Makowski, D., Aveline, A., Girardin, P., & Plantureux, S. (2008). Agri-environmental indicators to assess cropping and farming systems: A review. *Agronomy for Sustainable Development, 28*(1), 139–149.

Bohanec, M. (2011). DEXi: A program for multi-attribute decision making. http://www-ai.ijs.si/MarkoBohanec/dexi.html. Accessed 4 January 2012.

Bohanec, M., Messean, A., Scatasta, S., Angevin, F., Griffiths, B., Krogh, P. H., Znidarsic, M., & Dzeroski, S. (2008). A qualitative multi-attribute model for economic and ecological assessment of genetically modified crops. *Ecological Modelling, 215*(1–3), 247–261.

Bravin, E., & Kilchenmann, A. (2010). Ländervergleich der Apfelproduktion (International comparison in apple production). *Agrarforschung Schweiz, 1*(2), 52–59.

Castoldi, N., Bechini, L., Steinberger, J., Binder, C. (2007). Sustainability solution space using agro-ecological indicators at field level. Presentation held at Farming Systems Design 2007, Catania, Italy, September 10–12.

ENDURE. (2009). Interview with Pierre Ricci ENDURE coordinator. http://www.endure-network.eu/about_endure/all_the_news/interview_pierre_ricci_endure_coordinator. Accessed 4 January 2012.

Foster, C., Green, K., Bleda, M., Dewick, P., Evans, B., Flynn, A., Mylan, J. (2006). Environmental impacts of food production and consumption: A report to the Department for Environment, Food and Rural Affairs. Manchester Business School. Defra, London.

Frischknecht, R., Jungbluth, N., Althaus, H. -J., Doka, G., Dones, G., Heck, T., Hellweg, S., Hischier, R., Nemecek, T., Rebitzer, G., Spielmann, M., Wernet, G. (2007). Overview and methodology. Swiss Centre for Life Cycle Inventories. Dübendorf, ecoinvent report.

Gaillard, G., & Nemecek, T. (2009). Swiss Agricultural Life Cycle Assessment (SALCA): An integrated environmental assessment concept for agriculture. In: International Conference "Integrated Assessment of Agriculture and Sustainable Development, Setting the Agenda for Science and Policy", Egmond aan Zee, The Netherlands. AgSAP Office, Wageningen University, 134–135.

Grenz, J., Thalmann, C., Stämpfli, A., Studer, C., Häni, F. (2009). RISE – A method for assessing the sustainability of agricultural production at farm level. Rural Development News 1/2009.

Guinée, J. B., Gorrée, M., Heijungs, R., Huppes, G., Kleijn, R., de Koning, A., van Oers, L., Wegener Sleeswijk, A., Suh, S., Udo de Haes, H. A., de Bruijn, H., van Duin, R., Huijbregts, M. A. J., Lindeijer, E., Roorda, A. A. H., & Weidema, B. P. (2001). *Life cycle assessment – An operational guide to the ISO standards*. Den Haag/Leiden: Ministry of Housing/Spatial Planning and Environment (VROM) and Centre of Environmental Science (CML).

Gutsche, V., & Strassemeyer, J. (2007). SYNOPS – ein Modell zur Bewertung des Umwelt-Risikopotentials von chemischen Pflanzenschutzmitteln. Institut für Folgenabschätzung im Pflanzenschutz, Biologische Bundesanstalt für Land- und Forstwirtschaft.

HAIR. (2007). GIS related database. http://www.rivm.nl/rvs/risbeoor/Modellen/Results_of_HAIR_project.jsp. Accessed 4 January 2012.

Hauschild, M. Z., & Wenzel, H. (1998). *Environmental assessment of products* (Vol. 2: Scientific background. 2). London: Chapman & Hall. 565 p.

Hischier, R., Weidema, B., Althaus, H. -J., Bauer, C., Doka, G., Dones, R., Frischknecht, R., Hellweg, S., Humbert, S., Jungbluth, N., Köllner, T., Loerinick, Y., Margni, M., Nemecek, T. (2009). Implementation of life cycle impact assessment methods, Datav2.1. Swiss Centre for Life Cycle Inventories, Dübendorf, ecoinvent report.

IPCC. (2006). IPCC guidelines for national greenhouse gas inventories. In S. Eggleston, L. Buendia, K. Miwa, T. Ngara, & K. Tanabe (Eds.), *Agriculture, forestry and other land use* (Vol. 4). Kanagawa: IPCC.

ISO. (2006). ISO 14044, 1st Edition, – Environmental management – Life cycle assessment – Requirements and guidelines. Geneva, 46 p.

Lô-Pelzer, E., Bockstaller, C., Lamine, C., Mouron, P., Messéan, A. (2009). DEXi-PM: A model for ex ante sustainability assessment of innovative crop protection strategies. Presentation held at Farming System Design, Monterey (USA), 23–26 August 2009. http://www.endure-network.eu/about_endure/all_the_news/dexipm_assessing_cropping_system_sustainability. Accessed 4 January 2012.

Mila i Canals, L., Cowell, S. J., Sim, S., & Basson, L. (2007). Comparing domestic versus imported apples: A focus on energy use. *Environmental Science and Pollution Research, 14*(5), 338–344.

Mouron, P., & Scholz, R. W. (2007). Management influence on income risk in an apple production system on Swiss fruit farms. *International Journal of Fruit Science, 7*(4), 47–70.

Mouron, P., Nemecek, T., Scholz, R. W., & Weber, O. (2006a). Management influence on environmental impacts in an apple production system on Swiss fruit farms: Combining life cycle assessment with statistical risk assessment. *Agriculture, Ecosystems & Environment, 114*(2–4), 311–322.

Mouron, P., Scholz, R. W., Nemecek, T., & Weber, O. (2006b). Life cycle management on Swiss fruit farms: Relating environmental and income indicators for apple-growing. *Ecological Economics, 58*(3), 561–578.

Nemecek, T., & Erzinger, S. (2005). Modelling representative life cycle inventories for Swiss arable crops. *International of Journal of Life Cycle Assessment, 10*(1), 68–76.

Nemecek, T., & Kägi, T. (2007). Life cycle inventories of Swiss and European agricultural production systems. Final report ecoinvent V2.0 No. 15a, Agroscope Reckenholz-Taenikon Research Station ART, Swiss Centre for Life Cycle Inventories, Dübendorf, ecoinvent report.

Nemecek, T., Huguenin-Elie, O., Dubois, D., Gaillard, G. (2005). Ökobilanzierung von Anbausystemen im schweizerischen Acker- und Futterbau. Schriftenreihe Nummer 58, Agroscope ART Zurich, pp. 53–59.

Nemecek, T., von Richthofen, J.-S., Dubois, G., Casta, P., Charles, R., & Pahl, H. (2008). Environmental impacts of introducing grain legumes into European crop rotations. *European Journal of Agronomy, 28*(3), 380–393.

Sadok, W., Angevin, F., Bergez, J. E., Bockstaller, C., Colomb, B., Guichard, L., Reau, R., Messéan, A., & Dore, T. (2009). MASC, a qualitative multi-attribute decision model for ex ante assessment of the sustainability of cropping systems. *Agronomy for Sustainable Development, 29*(3), 447–461.

Udo de Haes, H., & Lindeijer, E. (2002). The conceptual structure of life-cycle impact assessment. In: H. Udo de Haes, G. Finnveden, M. Goedkoop, M. Hauschild, E. G. Hertwich, P. Hofstetter, W. Klöpffer, W. Krewitt, E. W. Lindeijer, R. Mueller- Wenk, S. I. Olson, D. W. Pennington, J. Potting, B. Steen (eds.), Life-cycle impact assessment: Striving towards best practice. SETAC Press, pp. 209–225.

Wiek, A., & Binder, C. (2005). Solution spaces for decision-making – A sustainability assessment tool for city-regions. *Environmental Impact assessment Review, 25*, 589–608.

Part IV
Sustainability Assessment in Organic and Multifunctional Systems

Chapter 10
Converting Suckler Cattle Farming Systems to Organic Farming: A Method to Assess the Productive, Environmental and Economic Impacts

Patrick Veysset, Michel Lherm, and Didier Bébin

Abstract This chapter proposes a method for assessing the farming system adaptations required in converting to organic farming (OF) in three beef production systems employed in the Charolais area. The conversion to OF was simulated by coupling an economic optimisation model ("Opt'INRA") with a model assessing non-renewable energy (NRE) consumption and greenhouse gas emissions ("PLANETE"). After adaptation of the production system, meat production decreased by 19–37%, depending on the initial level of intensification. The reduced use of inputs results in a 23–45% drop in non-renewable energy consumption per hectare and a 5–16% drop per ton of live weight produced. The shift to OF does significantly not affect gross GHG emissions per ton of live weight produced, but, taking into account carbon sequestration in grasslands, net GHG emissions could be lower for OF systems. Economically, the drop in productivity is not compensated by the gain in the meat selling price (+5% to +10%), gross farm product drops by 9–16%, and the lower use of inputs entails a strong drop in operational costs: –9% to –52%. Farm income falls more than 20% (–7% to –46%).

Keywords Beef production • Organic farming • Greenhouse gas • Non-renewable energy • Economics • Farm model

10.1 Introduction

Cattle farming has been singled out as a global-scale driver of global warming due to the levels of greenhouse gas (GHG) emissions generated. According to the French GHG emissions inventory (CITEPA 2009), in 2007, the agriculture sector was

P. Veysset (✉) • M. Lherm • D. Bébin
INRA, UR1213 Herbivores, site de Theix, F-63122, Saint-Genès-Champanelle, France
e-mail: veysset@clermont.inra.fr; lherm@clermont.inra.fr; bebin@clermont.inra.fr

A.A. Marta-Costa and E. Silva (eds.), *Methods and Procedures for Building Sustainable Farming Systems: Application in the European Context*,
DOI 10.1007/978-94-007-5003-6_10,

responsible for 19% of the national global warming potential (GWP), with livestock farming accounting for 53% of agriculture-related GHG emissions. De Vries and De Boer (2010) showed that the production of one kg of beef used more land and energy and had a higher global warming potential than the production of one kg of pork, chicken, egg or milk from non-organic production systems. Neither inventory figures nor these results take soil carbon sequestration into account.

In farming, productive and economic performance assessments and environmental performance assessments are inseparably linked. Various different LP models have already been developed for assessing the economic and environmental performance of farming systems (Janssen and Van Ittertum 2007), several of which have assessed the ecological and/or economic benefits of switching to organic farming (Benoit and Veysset 2003; Pacini et al. 2004; Kerselaers et al. 2007). The few models focused on beef production were developed to assess fodder systems and alternative feed regimens (Nielsen and Kristensen 2007).

Organic farming (OF) is perceived to be a more sustainable production system, performing better than conventional farming in terms of nitrogen losses, pesticide risk, herbaceous plant biodiversity (Pacini et al. 2003; Bengtsson et al. 2005) and water use (Wood et al. 2006). However, the main environmental issues that beef farmers in grassland areas have to contend with due to mounting societal pressure are GHG emissions and energy consumption. A number of recent studies have focused on the assessment of beef production GHG emissions (Phetteplace et al. 2001; Casey and Holden 2006a; Vergé et al. 2008; Beauchemin et al. 2010; Pelletier et al. 2010), including a comparison between organic and conventional production systems (Casey and Holden 2006b), but none of these studies has connected environmental results with economic results at farm scale.

The main objective of this study was to assess the farming system adaptations required in converting to OF, together with the consequences in terms of economic performance and the impacts on non-renewable energy (NRE) consumption and greenhouse gas emissions in three beef production systems employed in the Charolais area.

Given this objective, we opted to use a linear programming-based optimisation model. To study the revenue-maximising production system adaptations required in response to the conversion to OF and to assess economic and environmental performance, we used models coupled by Veysset et al. (2010) to account for suckler farming system diversity: (i) the economic optimisation model and (ii) an environmental assessment model.

10.2 Materials and Methods

10.2.1 The Models Used

We modelled and assessed the different farming systems in the Charolais area using the Opt'INRA Charolais systems optimisation model (Veysset et al. 2005). This model, which was built by linear programming in Excel, optimises the production system in order to maximise the gross profit margin for Charolais-based mixed crop-livestock farms running either calf-to-weanling systems (suckler cow farms rearing progeny to 9–18 months and selling to specialised fatteners, especially in Italy) or calf-to-beef systems (suckler cow farms rearing progeny to slaughter). Opt'INRA integrates all the animal and crop farming activities found in the Charolais basin, together with the different CAP and agro-environmental premiums with their allocation rules. All the activities are linked and bounded by different constraints: structural (useable area, arable land, etc.), agronomic (cropping plan, previous use, etc.), zootechnical (replacement rate, mortality and numerical productivity, feed requirements, etc.) and administrative (premiums). The objective function of the model is to maximise gross profit margin:

Gross profit margin = animal sales + grain sales + subsidies (crop, set-aside, animals, agro-environmental, other CAP premiums) – variable animal costs (feed, veterinary, herd costs) – variable forage and crop costs (seeds, fertilisers, pesticides, harvest).

Opt'INRA determines the combination of activities that meet all the constraints and maximise gross profit margin.

The optimal farming system derived from Opt'INRA requires data input on the equipment base and on the number of hours in use for each item of equipment. Each farm task (ploughing, tillage, seeding, harvest, product handling, etc.) is covered by a list of various items of equipment of different size and power and for which decision rules are given. The model therefore selects the equipment needed for the system defined. Likewise, a specific spreadsheet calculates the building floorspace needed depending on the number and type of livestock to be housed and the fodder and farm equipment to be stocked. The calculation-decision rules account for the unit needs of each type of livestock according to whether they are housed in a stall barn or a loose-housing system. Each item of equipment and each building is characterised in terms of its per-hour or per-hectare energy consumption, together with their annual operating costs.

We used the PLANETE model (Bochu 2002) to assess the non-renewable energy requirements of the defined farming system and quantify its GHG emissions. PLANETE works at farm scale. It compiles and records all the direct and indirect NRE consumption and GHG emissions tied to the farming activity "from cradle to farm gate". The fate of the products "from farm gate to grave" is not included in this assessment.

We distinguish between the consumption of (i) direct NRE, such as petroleum products and electricity consumed on-farm, and (ii) indirect NRE consumed off-farm in producing and transporting the farm's factors of production, including farm equipment and farm buildings. The energy values are expressed in megajoules (MJ).

Methane (CH_4), nitrous oxide (N_2O) and carbon dioxide (CO_2) are the main GHGs emitted by farming. The contribution of each of these three GHGs is captured via an indicator called the global warming potential (GWP), expressed in tons of equivalent CO_2 (tCO_2eq), which is the sum of each of the gases weighted by their respective coefficients of equivalence, i.e. 1 t CO_2 = 1, 1 t CH_4 = 25 and 1 t N_2O = 298 (IPCC 2007).

PLANETE is a model built in Excel, making it easy to couple with Opt'INRA (Figure 10.1). The inputs to PLANETE are the outputs of the Opt'INRA model, i.e. number of calvings, type and number of animals produced, cropping plan, type and

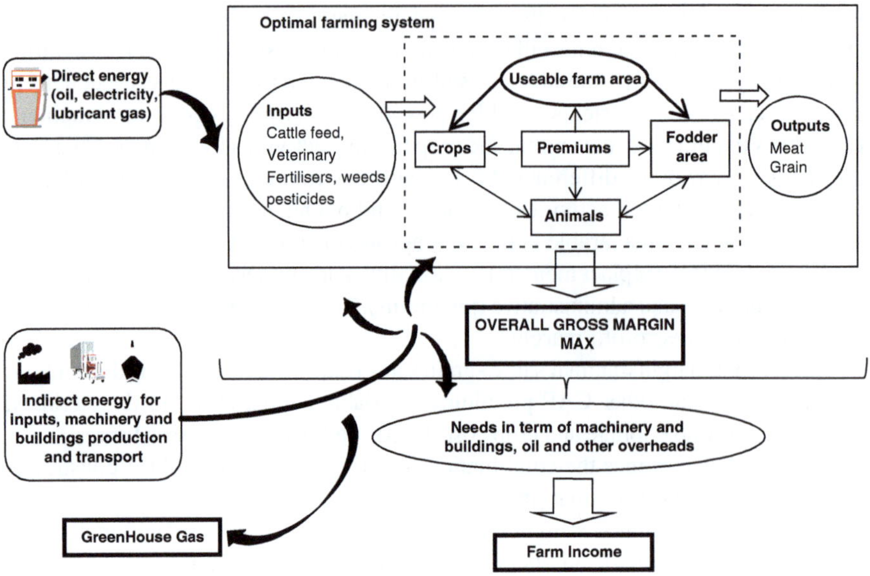

Fig. 10.1 Simplified diagram showing the coupling between an optimisation of the farming system maximising gross margin and a model assessing NRE consumption and GHG emissions

quantity of forages harvested and grazed, annual crop area, quantity of grain consumed on-farm and sold annually, feed ration per animal, quantity of forage and concentrate allocated, feed purchased and inputs purchased.

The description of the models and the methodology used for coupling has been fully described by Veysset et al. (2010).

10.2.2 GHG Emissions and Carbon Sequestration

PLANETE allowed us to determine gross GHG emissions, but carbon sequestration in grasslands (Soussana et al. 2007) is not taken into account.

The vegetation synthesises organic matter from CO_2 taken out of the atmosphere (photosynthesis and solar energy). A significant fraction of the biomass produced is incorporated into the soil (leaves, roots), thus enriching the soil with organic matter (OM) and thereby enhancing carbon (C) storage. However, this OM undergoes biotransformation (decomposition, mineralisation) that returns C to the atmosphere. Soil C storage/release dynamics will therefore depend primarily on changes in land use. According to Arrouays et al. (2002), carbon sequestration ranges from 200 kg/ha/year for a permanent pasture over 30 years old to 500 kg/ha/year for permanent pasture under 30 years old and temporary pasture. Given that we cannot know the exact age of permanent farm pastures, and given the broad difference between 200 and 500 kg/ha/year, we opted to work with an average 350 kg/ha/year to account for carbon sequestration in all permanent pastures. For temporary pasture, we also take into account C export when the pasture is ploughed, i.e. at 1,000 kg/ha/year.

Table 10.1 Carbon and CO_2 storage in pastures at farm level

	C t/year	CO_2 t/year
Permanent pastures (PP)	ha PP * 0.35	C * 44/12
Temporary pastures (TP)	(ha TP*0.50) – (ha TP ploughed*1.00)	

Based on the cropping plan and the proportion of temporary pasture ploughed each year, we can calculate, for the optimal system, the amount of C and CO_2 stored for 1 year (Table 10.1). This offsetting of GHG emissions was added to the model, after which net GHG emissions were calculated (Dollé et al. 2009).

10.2.3 Nitrogen Balance

Charolais-area beef farms conventionally use little chemical nitrogen fertiliser, averaging 40–50 kg N per ha of farm area per year. Nitrogen balance is low (+40 kg N/ha) and without fertilisers could be negative (although unsustainable for plant nutrition).

Since plant nitrogen supply is a key factor in OF, we integrated an N-balance constraint into the Opt'INRA model. N balance is the difference between nitrogen leaving the farm with the sale of beef and crop products and nitrogen entering the farm, including N fixed by legumes (Simon and Le Corre 1992). N balance had to be slightly positive (+30 kg N/ha) to counteract the losses due to run-off and volatilisation. The model can balance this N balance, either by buying in organic N fertiliser (a unit of organic N costs about 10 times more than a unit of chemical N) or by choosing an optimal cropping plan including protein-rich plants (Veysset 2002).

10.2.4 The Three Farms Studied

Charolais-based suckler cow farming is typified by a characteristically diverse range of farming systems: diverse land-use patterns (proportion assigned to cash crops) and a diverse mix of intensive-extensive agriculture, as well as a diverse range of animals raised (young or old, lean or fat). The Réseaux d'Élevage system run and supervised by the Institut de l'Élevage has pinpointed 14 specialised Charolais beef production systems that are studied as test cases (Réseau d'Élevage 2006). Each test case is a farm model operating at established steady-state rate. We selected three specialised beef-producer test cases for study – not for their representativeness but for the diverse range of systems they covered:

- A: calf-to-weanling system producing male and female store cattle (13–16 months old) for the Italian market. This test case A holds 68.4 suckler cow premium entitlements, and the buildings have capacity for 77 cows. 100% of 100 ha of the total farm area is devoted to grass; there is no arable land.

- B: calf-to-beef system producing beef steers over 30 months old. 55% of males are castrated, while the others are sold as weaners. Heifers (30–36 months old) and cull cows are fattened on-farm. This test case B holds 60.8 suckler cow premium entitlements, and the buildings have capacity for 68 cows. Total farm area is 125 ha, of which 90.4 ha (72%) is under permanent grass.
- C: calf-to-beef system producing young bulls (17 months old) and fattened heifers (27 months old). This is an intensive production system using maize silage (stocking rate, 1.60 LU/ha fodder area). This test case C holds 102.6 suckler cow premium entitlements, and the buildings have capacity for 110 cows. Total farm area is 155 ha, of which 100 ha (64%) is under permanent grass. The storage capacity for the maize silage is 130 tons of dry matter (approximately 12 ha).

Analyses of NRE consumption and GHG emissions were carried out with the optimised systems from Opt'INRA. Economic data (input and output prices) are the 4-year averages (2004–2008 excluding 2007). The baseline CAP setting is the 2003 Luxemburg reform. These optimised conventional systems gave a basis for comparisons with the OF systems for the same farms under the same economic and CAP conditions.

10.2.5 *Conversion to OF: Technical and Market Hypothesis*

Our long-term Charolais suckler cattle farms network, including more than 10% of OF-certified farms, makes it possible to analyse the average and time-course evolution of each zootechnical and economical dataset at the farm level (Veysset et al. 2009). Analysis showed that numerical productivity does not differ significantly between OF and conventional systems: annual pregnancy and calf mortality rates are similar between the two systems. However, under the more extensive production system, OF farmers use 50% less concentrate per livestock unit. Live weight of all the animals sold at the same age is 2–5% lower, and this difference remains the same over the years. These observations are in line with data on other breeds in other parts of France (Pavie and Lafeuille 2009).

In the conventional systems, more than 65% of males sold are 10- to 13-month-old weaners, these males being sold as store cattle to specialised fatteners in Italy. The remaining males are mainly sold fattened as baby beef (16–19 months old), and only 3% of the males from the conventional systems are sold as beef steers (30–36 months old). Most of the cull cows and heifers (27–36 months old) are fattened and sold for the French market. The OF farmers also sell more than 60% of their males as weaners for the Italian market. Since the OF specifications limit the indoor fattening period to 3 months and at most 1/5 of their lifetime, the production of fattened baby beef and heifers less than 30 months of age is not compatible with an OF production system.

There is no specific market for OF store animals. The OF-certified weaners (males and females) are sold in the same market as conventional weaners and are valued at conventional farm gate prices. Since OF weaners are lighter (less concentrates) and their body conformation is not as good, the price per kg live weight of the OF weaners

Table 10.2 Animals' weight, yields, inputs and output prices used for the conventional and OF systems

		Conventional	Organic farming
Animals sold			
Weaner males 10 months old	Kg live weight*€/kg lw	380*2.45	350*2.35
Weaner males 13 months old	Kg live weight*€/kg lw	473*2.35	406*2.40
Young store bulls 16 months old	Kg live weight*€/kg lw	490*2.35	436*2.40
Baby beef 17 months old	Kg carcass*€/kg cc	429*3.25	Not possible
Beef steers 28 months old	Kg carcass*€/kg cc	452*3.45	Not possible
Beef steers 31 months old	Kg carcass*€/kg cc	476*3.55	430*3.75
Beef steers 36 months old	Kg carcass*€/kg cc	482*3.58	442*3.80
Store cull cows	Kg live weight*€/kg lw	680*1.70	650*1.60
Fattened cull cows	Kg carcass*€/kg cc	435*3.30	405*3.65
Weaner females 8 months old	Kg live weight*€/kg lw	277*2.35	270*2.20
Weaner females 13 months old	Kg live weight*€/kg lw	353*2.25	331*2.20
Store heifers 16 months old	Kg live weight*€/kg lw	418*2.15	393*2.10
Beef heifers 27 months old	Kg carcass*€/kg cc	371*3.55	Not possible
Beef heifers 31 months old	Kg carcass*€/kg cc	378*3.65	370*3.80
Beef heifers 36 months old	Kg carcass*€/kg cc	394*3.75	385*3.90
Crops sold			
Cereals	t/ha*€/t	5.5*106	2.9*210
Pasture yield			
Permanent pasture	t dry matter/ha	5.35 to 7.85	4.30 to 6.30
Temporary pasture	t dry matter/ha	5.10 to 5.70	3.30 to 4.30
Purchased feed and fertilisers			
Soya meal	€/t	300	700
Cereals	€/t	200	390
Nitrogen units	€/kg	0.75 (Chemical)	7.00 (Organic)

can be kept lower than conventionally reared weaners. Only fattened cull cows, fattened steers and heifers over 30 months old are valued as OF products. The average premium price observed over the last 4 years is only +5% to +10% per kg carcass for the OF-fattened animals. We allowed for the possibility of castrating males even on holdings where this had never been previous practice. Castration was limited to 50% of the males of each generation, which is the level usually found among livestock farmers.

The ban on chemical fertilisers results in a drop in 15–25% in pasture yield and a 50% drop in cereal yield. The observed prices of the main OF-grade inputs (feeds cereals and soya meal, nitrogen units) are two to ten times higher than for conventional and/or synthetic inputs.

Table 10.2 summarises the weights of the animals and the yields and prices of the main inputs and outputs used in this study to calibrate our model, for both conventional systems and OF projections.

Opt'INRA was applied in a single-period approach. We started from a stable conventional situation and arrived at a stable OF situation. The conversion period was not taken into account. The only factors changing are production constraints, inputs and farm product prices. Farm structure (size, labour force and right to produce) was considered a constant.

10.2.6 Analysis, Expression of Results

The farm-scale energy audit covers all NRE consumption over 1 year, expressed in MJ, relative to the production of 1 ton (1,000 kg) of live weight (LW) over 1 year (MJ/t LW). We also calculated NRE consumption per ha of farmland devoted to cattle production (ha "bovine" = fodder area + home-consumed cereals).

In the same way, GHG emissions are expressed in tons of equivalent CO_2 per ton of live weight produced (tCO_2eq/t LW) and per ha of farmland devoted to cattle production.

Taking the gross profit margin generated by Opt'INRA, deducting the specific mechanisation and building costs (including depreciation costs) and other fixed costs (independent of the farm system deployed) gives the farm income of the test case studied.

10.3 Results

10.3.1 Production Systems and Farm Income

Table 10.3 gives the technical and economic results of the three farms studied for conventional and OF systems. These results are the Opt'INRA outputs.

Farm A: As this all-grass system does not produce its own concentrates, it has to buy them if it wants to fatten some animals. The model chooses to fatten all the heifers but only 30% of the cull cows. The males are all sold as weaners. Calvings decrease by 23% and stocking rate decreases 10%. Live weight produced drops 19% and gross farm product decreases 12%. The quantity of concentrates per LU decreases by 25%, but due to price increases in purchased concentrates, the concentrates' cost per LU increases by 33% (140 €/LU vs 105). The lower number of LU and the savings in chemical fertilisers makes it possible to cut operational costs by 9%. However, farm income suffered the highest drop of the three test case scenarios: –46%, i.e. –94 €/ha.

Farm B: The adaptations made consist in decreasing the number of calvings (−15%) and producing younger beef heifers (31 months old) to adapt stocking rate. The kg of live weight produced decreases by 19%. Opt'INRA chooses to reallocate grassland over to maize silage and grain; the OF system can decrease the quantity of concentrates per LU by 43%. This system becomes feed self-sufficient. Operational costs decrease by 40%, while gross farm product only decreases 9% and overall gross margin close to that of the conventional system. However, due to the higher area under grain and maize, mechanisation costs are increased 3%. Farm income decreases by 7%, i.e. −16 €/ha.

Farm C: Since this farm is the most intensive conventional system, the shift to OF entails the highest drop in number of calvings (–25%) and stocking rate (–24%).

Table 10.3 Opt'INRA outputs: technical and economic results of the 3 farms studied for conventional (Conv.) and OF systems

	A: calf-to-weanling 100% grassland farm		B: calf-to-beef. Beef steers production		C: calf-to-beef. Intensive baby beef production	
	Conv.	OF	Conv.	OF	Conv.	OF
Total farm area, ha	100	100	125	125	155	155
Cash crops, ha	0.0	=	0.0	0.0	0.0	4.1
Grain home-consumed, ha	0.0	=	15.8	16.6	35.8	18.0
Fodder area, ha	100.0	=	109.2	108.4	119.2	132.9
Including maize silage	0.0	=	1.8	7.1	11.0	9.8
Hay + grass silage, ha	45.4	52.6	54.0	54.5	42.4	72.5
Number of calvings	73	56	68	58	107	80
Livestock units, LU	107.4	95.9	130.6	112.6	192.6	142.3
Stocking rate LU/ha "bovine"[a]	1.07	0.96	1.04	0.90	1.24	0.94
Males sold[b]	W13	W16	W13 + Bs31	=	Bb17	W13 + Bs31
Heifers sold[c]	Sh16	Bh31	Bh36	Bh31	Bh27	Bh31
% Cull cows fattened	30	=	100	=	100	100
Concentrates, kg/LU	473	357	743	422	1 248	374
Including purchased,%	100	100	11	0	14	0
Live weight (LW) produced, kg	35,091	28,404	41,268	33,584	69,334	43,941
LW produced kg/ha "bovine"[a]	351	284	330	269	447	291
Grain sold, t	–	–	–	–	–	12.2
Nitrogen balance, kg N/ha	54	42	58	30	77	30
Bovine gross margin, €/LU	607	549	604	623	602	686
Fodder area gross margin, €/ha	673	571	716	676	894	743
Crop gross margin, €/ha	–	–	–	–	–	517
Total product, €	110,853	97,954	132,646	120,309	199,422	167,008
Operational costs, €	30,027	27,376	27,745	16,740	52,506	24,967
Overall gross margin, €	80,826	70,579	104,901	103,569	146,915	142,041
Fixed costs, €	60,143	59,318	76,480	77,121	99,123	99,023

(continued)

Table 10.3 (continued)

	A: calf-to-weanling 100% grassland farm		B: calf-to-beef. Beef steers production		C: calf-to-beef. Intensive baby beef production	
	Conv.	OF	Conv.	OF	Conv.	OF
Including mechanisation, €	20,843	20,018	27,890	28,621	32,783	32,683
Farm income, €	20,683	11,261	28,422	26,448	47,792	43,019

[a]ha "bovine": total area devoted to the herd = fodder area + cereals home-consumed
[b]Males sold: W10 = weaners 10-month-old, W13 = weaners 13-month-old, W16 = weaners 16-month-old, Bb17 = baby beef 17-month-old, Bs31 = beef steers 31-month-old, Bs36 = beef steers 36-month-old
[c]Females sold: W8 = weaners 8-month-old, Sh16 = store heifers 16-month-old, Bh27 = beef heifers 27-month-old, Bh31 = beef heifers 31-month-old, Bh36 = beef heifers 36-month-old

The OF system produces 31-month-old beef steers and heifers which are fattened mainly with grass. Concentrate requirements per LU decrease by 70% (374 kg/LU), and the area devoted to home-consumed cereals decreases by 50% (18 ha), while the area under grass increases by 14% (123.1 ha vs 108.2). The system can release 4.1 ha to cash crops. The total live weight produced decreases by 37%, but 12.2 tons of grain are sold. Farm product decreases by 16% and operational costs are cut by 52%. This farm undergoes the highest changes in the system, and overall, the drop in the farm income is 10% (–31 €/ha).

For all the farms where grain production is possible, a mixture of cereals/protein-rich plants is prioritised in order to supply nitrogen to the system (soil and animals). Unlike conventional systems, no OF systems exclusively grow one cereal only.

The cropping plan, inputs use and outputs for the OF systems respect the nitrogen balance constraint (+30 kg N/ha). This nitrogen balance is from 22% to 61% lower for the OF systems than for conventional systems, and the most intensive conventional system (farm C) makes the heaviest reduction in its nitrogen balance (–47 kg N/ha). Nitrogen leaching risk is thus lower for the OF systems than for conventional farms.

10.3.2 Non-renewable Energy Consumption

Table 10.4 gives the non-renewable energy consumed for beef production for conventional and OF systems.

10.3.2.1 Conventional Systems

NRE consumption required to produce 1 t of live weight ranges from 27,254 to 33,483 MJ. Petroleum products (fuel, lubricants) are the main consumption input,

Table 10.4 Non-renewable energy consumed for beef production for conventional (Conv.) and OF systems

	A: calf-to-weanling 100% grassland farm		B: calf-to-beef. Beef steers production		C: calf-to-beef. Intensive baby beef production	
	Conv.	OF	Conv.	OF	Conv.	OF
Direct energy (MJ/t LW)						
Fuel and lubricant	7,962	9,671	10,971	13,069	9,080	12,199
Electricity and water	2,320	2,471	2,196	2,369	1,766	1,915
Indirect energy (MJ/t LW)						
Purchased feed	4,351	3,430	1,059	105	1,813	293
Artificial fertilisers	4,261	648	7,559	621	7,906	451
Seeds and treatments	0	0	797	1,013	1,132	924
Veterinary and various raising inputs	1,952	2,018	1,826	2,018	1,628	2,051
Machinery	4,396	5,456	7,051	8,141	5,622	5,793
Buildings	2,011	2,107	2,024	2,116	1,807	2,196
Total MJ/t LW	27,254	25,801	33,483	29,452	30,755	25,821
Total MJ/ha "bovine"	9,564	7,329	11,054	7,913	13,757	7,518

responsible for around 30% of total NRE consumption. The second-highest consumption input is fertilisers and soil improvers, at 16–26% of total NRE consumption, almost level with farm equipment inputs (NRE used in manufacturing and delivering the farm equipment). Feed purchases account for only 3–16% of NRE consumption. Miscellaneous other process procurements required for farming livestock (veterinary products, salt/minerals) and harvesting fodder (plastic bale wraps, strings) account for 6–7% of total NRE consumption. Depreciation of the energy required to build the farm buildings represents only 57% of NRE consumption. Finally, the list is rounded up by purchases of seed and plant protection agents, which account for only 0–2.5% of the NRE consumed to produce 1 ton of live weight.

The most energy-efficient beef production farm was the all-grass system (test case A). This farm was forced to buy in all its feed (4,351 MJ/t LW) but demonstrated some of the lowest consumption levels for fuel, fertiliser and equipment (7,962, 4,261 and 4,396 MJ/t LW, respectively).

The most intensive system (farm C) was the second-most energy-efficient beef production farm. Fertiliser and feed purchases account for a high proportion of NRE consumption, but the weights of the fuel, mechanisation and building factors are lower than in the other systems.

Mechanisation, which goes in tandem with fuel, is the primary source of variability in results on NRE consumption for 1 ton of live weight produced, with feed purchases coming second.

Crossing the data against farm area used by the herd reveals a direct link between farm intensification level (kg LW produced/ha "bovine") and NRE consumption, which varies from 9,564 (farm A) to 13,757 (farm C) MJ/ha "bovine". The leading influencing factor remains fuel, with fertilisers ranking second.

10.3.2.2 Organic Farming Systems

This reduced use of allied crop input (fertilisers, seed, treatments) under organic systems, especially the non-use of chemical fertilisers, leads to a 60–70% drop in the consumption of NRE/t LW related to these items.

Farms B and C are almost completely self-sufficient on animal feed, as only minerals are bought in. Consumption of NRE related to the purchase of food therefore drops almost 90%.

At constant surface area, organic and conventional systems share almost identical equipment requirements. As total live weight production drops, the mechanisation/t LW item increases by 18%, 13% and 8% for farms A, B and C, respectively. The direct energy/t LW item increases 17% (farm B) and 30% (farm C).

All in all, the shift to organic farming entails a significant decrease (–5% to –20%) in the consumption of NRE/t LW produced, under all systems. This fall in consumption of NRE/t LW was only 5% for the all-grass system (farm A), which is less intensive and uses less inputs than the conventional system.

Since OF systems are less intensive (stocking rate –10% to –24%), they use far fewer inputs per ha "bovine". The consumption of direct energy per ha thus shows a 5–15% decrease, while consumption of NRE per ha linked to crop inputs shows a 65–80% decrease. Total NRE consumption per ha "bovine" is 23–45% lower in OF systems than conventional systems.

10.3.3 GHG Emissions and Carbon Sequestration

Table 10.5 gives the greenhouse gas emission for beef production for conventional and OF systems.

10.3.3.1 Conventional Systems

The conventional Charolais suckler cattle farms systems produce 14.9–17.2 tCO_2eq/ton of live weight produced over 1 year and 5.58–6.68 tCO_2eq/ha "bovine".

Methane emissions tied exclusively to ruminant farming (enteric fermentation and manure management) are the main driver of gross GHG emissions, at around 60% (from 58% to 66%). Ruminant activities are also responsible for over 50% of farm N_2O emissions, principally from urine and faecal waste at pasture. Livestock is responsible for nearly 75% of farm-scale gross emissions (69–80%), followed by farm inputs, especially mineral fertilisers which alone account for 5–11%. The combustion of direct energy sources (fuel and electricity) accounts for 27% of CO_2 emissions but only 4% of farm-scale GWP.

Cows are the biggest driver of GHG emissions in the herd. Calf-to-weanling systems producing store animals, where cows account for the largest share of herd livestock units, have a higher GWP per ton of live weight or per hectare than calf-to-beef

Table 10.5 Greenhouse gas emissions for beef production for conventional (Conv.) and OF systems

	A: calf-to-weanling 100% grassland farm		B: calf-to-beef. Beef steers production		C: calf-to-beef. Intensive baby beef production	
	Conv.	OF	Conv.	OF	Conv.	OF
CO_2 (tCO_2eq/t LW)	1.9	2.0	2.5	2.5	2.2	2.3
Combustion of direct energy	0.5	0.6	0.7	0.8	0.6	0.8
Inputs making	1.4	1.3	1.8	1.6	1.6	1.5
CH_4 (tCO_2eq/t LW)	11.5	11.4	9.8	10.5	8.8	10.6
N_2O (tCO_2eq/t LW)	3.9	3.6	4.7	3.8	3.9	3.8
Inputs making	0.2	0.0	0.3	0.0	0.4	0.0
Nitrogen application on farm area	1.4	1.1	2.0	1.3	1.6	1.4
Cattle waste	2.3	2.5	2.3	2.4	2.0	2.4
Gross GHG emissions (tCO_2eq/t LW)	17.2	17.0	16.9	16.7	14.9	16.7
Gross GHG emissions (tCO_2eq/ha)	6.05	4.83	5.58	4.48	6.68	4.85
Carbon storage t	35.0	35.0	36.7	34.9	37.5	41.9
C offset% gross GHG emissions	21	27	19	23	13	21
Net GHG emissions (tCO_2eq/t LW)	13.6	12.5	13.6	12.9	12.9	13.2
Net GHG emissions (tCO_2eq/ha)	4.77	3.55	4.50	3.46	5.79	3.83

systems and at practically identical stocking rates. The least GHG-emitting farm in terms of ton of LW produced is therefore C, where all animals are fattened and where cows account for 48% of LU. Gross GHG emissions are 14.9 tCO_2eq/t LW for C, with CH_4 representing 59% of these emissions. Farm A, which sells most of its animals as store cattle and where cows represent 57% of total LU, generates 17.2 tCO_2eq/t LW.

Stocking rate (number of animals raised and produced per hectare and thus quantity of live weight produced per hectare) is the main driver of herd-related GHG emissions per ha. The most intensive test case C (1.24 LU/ha "bovine") is the most gross GHG-emitting farm per ha. With its lowest stocking rate (1.04 LU/ha "bovine") and calf-to-beef system, B is the lowest gross GHG-emitting farm per ha for cattle production.

Depending on the share split of permanent and temporary pastures in the total farm area, and thus on the ha of grassland per t LW produced, the carbon offset can be more or less important. With farms A, B and C producing 2.85, 2.60 and 1.57 ha of grassland/ton LW produced, the offsetting of gross GHG emissions/t LW is 21%, 19% and 13%, respectively. Net GHG emissions ranged from 12.9 (farm C) to 13.6 (farms A and B) tCO_2eq/t LW and from 4.50 (farm B) to 5.79 (farm C) tCO_2eq/ha "bovine".

Farm C was the lowest gross GHG-emitting farm, but as it contained a lower proportion of grassland, the live weight was produced with more grain and maize than in the other farms, with the result that it showed the lowest carbon offset. Net GHG emissions remained the lowest but were only 5% lower than for farm A, while gross GHG emissions were 13% lower.

10.3.3.2 Organic Farming Systems

The shift to OF had no significant impact on gross GHG emissions/ton LW produced. As CH_4 is the main driver of GHG emissions, OF has no impact. Indeed, due to the lower live weight production/LU, gross GHG emissions/t LW produced can even prove higher under OF systems, at +12% for farm C.

Due to the lower pasture productivity and thus the lower stocking rates, OF systems use more ha of pastures to produce 1 ton of LW, i.e. +24% (3.52 ha/t LW), +16% (3.02 ha/t LW) and +78% (2.80 ha/t LW) for farms A, B and C, respectively. The offsetting of gross GHG emissions/t LW due to the carbon sequestration ranges from 21% (farm C) to 27% (farm A), i.e. 3.5–7.7 points higher than for conventional farming systems.

While gross GHG emissions from OF systems are on par with or even slightly higher than conventional systems, the net GHG emissions are the same (+2% for farm C) or from 5% to 8% lower for farms B and A, respectively.

As stocking rate is lower under OF systems than on conventional farms, CH_4 emissions per ha "bovine" ranged from 13% to 20% lower. This drop in CH_4 emissions coupled with the reduced use of inputs per ha "bovine" means that the shift to OF leads to a –20% to –27% cut in gross GHG emissions and a –23% to –34% cut in net GHG emissions.

10.4 Discussion

The energy content of animal biomass is measured calorimetrically as raw energy (or feed value). A ton of bovine live weight equates to 14,000 MJ. One ton of LW from a conventional Charolais-based suckler cow herd requires about 31,000 MJ of NRE to produce 14,000 MJ for food, whereas one ton of LW from an OF system requires about 27,000 MJ of NRE to produce this same 14,000 MJ of energy for food. The "energy efficiency" ratio is 0.45 MJ of NRE for 1 MJ of food energy from a conventional production system, compared to 0.52 (i.e. 13% better) for an OF system.

OF is a more energy-efficient system for beef production, as energy is saved on the non-use of chemical fertilisers and other inorganic industrially produced inputs. Similar patterns are reported for other agricultural products: OF systems use 10–20% less NRE to produce one ton of cereal than conventional systems (Dalgaard et al. 2001; Refsgaard et al. 1998; Bochu 2007) and 15–30% less NRE to produce 1,000 l of milk (Refsgaard et al. 1998; Cederberg and Mattsson 2000; Grönroos et al. 2006;

Bochu 2007). Based on 35 surveys led on farms in southern Germany, Haas et al. (2001) reported that organic farms used 55% less NRE to produce one ton of milk. However, the results on farming monogastric animals show a different pattern: NRE use per kg of pig produced is 40% higher in French OF systems (Basset-Mens and Van der Werf 2005), whereas for eggs and poultry production in the UK, NRE use per ton is 10% higher in organic systems (Azeez 2008). However, after analysis of 15 crop and livestock sectors weighted in relation to the UK's total agricultural output, switching these 15 sectors to organic farming would decrease total NRE consumption by 26% (Azeez 2008).

The impact of the conversion to OF on the GWP is not really significant per ton of live weight produced. Only a higher proportion of grassland in the farm area can make a difference, as it increases the carbon offset and thus decreases net GHG emissions. Without taking this carbon sequestration into account, Casey and Holden (2006b) reported that organic Irish suckler-beef units emit 14% less GHG/t LW. Cederberg and Mattson were unable to draw concrete conclusions on switching milk production to OF, whereas Hass et al. (2001) found the same levels of GHG emission/t milk in organic and intensive systems. GWP seems to be much higher for organic pig production than conventional pig farming at +70% (Basset-Mens and Van der Werf 2005).

As most of the conventional Charolais suckler cattle systems are grassland-based systems, the shift to OF does not improve carbon sequestration, and thus the organic matter balance, which remains at a good level. However, for the intensive mixed crop-livestock system, carbon sequestration can remain low, and since increasing the share of grasslands in the total farm area is one of the best ways to improve organic matter balance, OF livestock systems perform better on this criterion than conventional systems.

Most papers have used tons or kg produced as the main functional unit for analysing the results. In some cases, the surface area of agricultural land used for the production could be a useful scalar to analyse the results, especially where the preservation of abiotic resources (water, soil, etc.) is a major issue. Consumption of NRE and net GHG emissions per ha devoted to beef production are much lower under organic systems. These better per ha results for OF are mainly due to the lower stocking rate and thus to lower outputs per ha for organic systems but also to the better N efficiency (Olesen et al. 2006).

Farruggia et al. (2006) have shown that floristic diversity, both at farm level and field level, increases when the stocking rate decreases in suckler cattle systems on Massif Central grassland. Converting these systems to OF could improve biodiversity by making them less intensive with a lower stocking rate.

From an economics point of view, this lower productivity of organic systems is not totally compensated by cost savings on Charolais suckler cattle farms, and at constant structure, farm income can drop significantly enough to become unacceptable for farmers. One of the key points for farm economics is that there is no market for OF store cattle, whereas farmers prefer producing weaners instead of beef steers because they do not have to capitalise over more than 30 months, making it a less risky option in terms of cash flow. The economic results of this study are calculated based on average prices over 4 years. However, OF prices tend to remain stable, whereas conventional prices experience significant fluctuation, with the result that the price differential between organic and conventional products rose from 2% (2006) to 11% (2008) per kg carcass of beef steers and cull cows. With the 2008 prices, the farm income of OF farms will be similar or higher than for conventional farms. With our price hypothesis, for the four test cases studied in this chapter, the drop in farm income ranged from −16 Euros/ha (farm B) to −94 Euros/ha (farm A). If OF offers better per ha environmental performance, then switching to OF could be made financially viable. Given the twin global challenges of ensuring food security and reducing GHG emissions, the main policy imperative for decision makers is to explicitly combine these two goals (Garnett 2009). Under the Common Agricultural Policy "Health Check" system, France has decided to redirect subsidies towards grassland livestock production systems and sustainable farming (Ministère de l'Agriculture et de la Pêche 2009). In this context, OF-certified grassland areas are earmarked to receive aid amounting to 80–100 Euros/ha.

10.5 Conclusions

Improving energy efficiency, self-reliance and carbon footprint are some of many principles and objectives in organic agriculture. These objectives need to be balanced against other objectives, such as agricultural output and farm income (Niggli et al. 2008).

Assessments of farm production systems should be fully holistic – environmental assessments cannot be divorced from economic assessments. Reducing the environmental footprint of a cattle farm only makes sense if the farm is economically viable and thus able to run sustainably.

Depending on whether analysis is focused on production and market, resource protection or farm or territory-wide economy, the decision maker will give more weight to certain criteria but should not ignore the others when seeking the best compromise.

Whole-farm models coupling biophysical, economic and environmental models offer powerful tools for carrying out multicriteria analysis on the opportunity to switch to a new production system.

Global warming is a real problem, and beef production is a central contributor. However, the differences between intensive indoor systems and grass-based pasture systems mean they do not have the same impacts. In depressed grassland zones and highland areas where grass is the sole resource, livestock production is not in competition with other use demands of the agricultural area, and more importantly, it allows this land to be maintained under grass (carbon sequestration) and human activities. A methodological challenge for the future assessment of farming systems is to conduct multicriteria analyses that also integrate social aspects (Siciliano 2009).

References

Arrouays, D., Balesdent, J., Germon, J. C., Jayet, P. A., Soussana, J. F., Stengel, P. (2002). *Contribution à la lutte contre l'effet de serre. Stocker du carbone dans les sols agricoles de France?* Expertise scientifique collective. Rapport INRA 332 p.

Azeez, G. S. E. (2008). *The comparative energy efficiency of organic farming.* (International Conference on Organic Agriculture and Climate Change, Enita of Clermont, ABioDoc, France, April 17–18th.)

Basset-Mens, C., & Van der Werf, H. (2005). Scenario-based environmental assessment of farming systems: The case of pig production in France. *Agriculture, Ecosystems & Environment, 105*, 127–144.

Beauchemin, K. A., Janzen, H. H., Shannan, M. L., McAllister, T. A., & McGinn, S. M. (2010). Life cycle assessment of greenhouse gas emissions from beef production in Western Canada: A case study. *Agricultural Systems, 103*, 371–379.

Bengtsson, J., Ahnström, J., & Weibull, A. C. (2005). The effects of organic agriculture on biodiversity and abundance: A meta-analysis. *Journal of Applied Ecology, 42*(2), 261–269.

Benoit, M., & Veysset, P. (2003). Conversion of cattle and sheep suckler farming to organic farming: Adaptation of the farming system and its economic consequences. *Livestock Production Science, 80*, 141–152.

Bochu, J. L. (2002). PLANETE: *Méthode pour l'analyse énergétique des exploitations agricoles et l'évaluation des émissions de gaz à effet de serre*. (Colloque national: Quels diagnostics pour quelles actions agroenvironnementales? 10 & 11 octobre, Solagro, 68–80.)

Bochu, J. L. (2007). *Synthèse 2006 des bilans PLANETE. Consommation d'énergie et émissions des GES des exploitations agricoles ayant réalisé un bilan PLANETE.* Rapport final. ADEME, Ministère de l'Agriculture et de la Pêche.

Casey, J. W., & Holden, N. M. (2006a). Quantification of GHG emissions from suckler-beef production in Ireland. *Agricultural Systems, 90*, 79–98.

Casey, J. W., & Holden, N. M. (2006b). Greenhouse gas emissions from conventional, agri-environmental scheme, and organic Irish suckler-beef units. *Journal of Environmental Quality, 35*, 231–239.

Cederberg, C., & Mattsson, B. (2000). Life cycle assessment of milk production – A comparison of conventional and organic farming. *Journal of Cleaner Production, 8*, 49–60.

CITEPA. (2009). Inventaire des émissions de gaz à effet de serre en France au titre de la convention cadre des Nations Unies sur les changements climatiques. *Rapport national d'inventaire, CCNUCC/CRF,* 1196 p. http://www.citepa.org/publications/Inventaires.htm

Dalgaard, T., Halberg, N., & Porter, J. R. (2001). A model for fossil energy use in Danish agriculture used to compare organic and conventional farming. *Agriculture, Ecosystems & Environment, 87*, 51–65.

De Vries, M., De Boer, I. J. M. (2010). Comparing environmental impacts for livestock products: A review of life cycle assessments. *Livestock Science, 128*, 1–11.

Dollé, J. B., Gac, A., & Le Gall, A. (2009). L'empreinte carbone du lait et de la viande bovine. *Rencontres Recherches Ruminants, 16*, 233–236.

Farruggia, A., Dumont, B., Jouven, M., Baumont, R., Loiseau, P. (2006). Caractériser la diversité végétale à l'échelle de l'exploitation dans un système bovin allaitant du Massif central. *Fourrages, 188*, 477–494.

Garnett, T. (2009). Livestock-related greenhouse gas emissions: Impacts and options for policy makers. *Environmental Science & Policy, 12*, 491–503.

Grönroos, J., Seppälä, J., Voutilainen, P., Seuri, P., & Koikkalainen, K. (2006). Energy use in conventional and organic milk and rye bread production in Finland. *Agriculture, Ecosystems & Environment, 117*, 109–118.

Haas, G., Wetterich, F., & Köpke, U. (2001). Comparative intensive, extensified and organic grassland farming in southern Germany by process life cycle assessment. *Agriculture, Ecosystems & Environment, 83*, 43–53.

Intergovernmental Panel on Climate Change. (2007). 2006 IPCC guidelines for national greenhouse gas inventories. http://www.ipcc-nggip.iges.or.jp/public/2006gl/index.html

Janssen, S., & Van Ittersum, M. (2007). Assessing farm innovations and responses to policies: A review of bio-economic farm models. *Agricultural Systems, 94*, 622–636.

Kerselaers, E., De Cock, L., Lauwers, L., & Van Huylenbroeck, G. (2007). Modeling farm-level economic potential for conversion to organic farming. *Agricultural Systems, 94*, 671–682.

Ministère de l'Agriculture et de la Pêche. (2009). Bilan de santé de la PAC: mise en oeuvre. Pour une PAC préventive, juste et durable http://agriculture.gouv.fr/sections/presse/communiques/legitimite-nouvelle-pour/downloadFile/FichierAttache_1_f0/Bilandesantefrance.pdf?nocache=1235412961.28

Nielsen, B. K., & Kristensen, A. R. (2007). Optimal decisions in organic beef production from steers – Effects of criterion of optimality and price changes. *Livestock Science, 110*, 25–32.

Niggli, U., Slabe, A., Schmid, O., Halberg, N., Schlüter, M. (2008). Vision for an organic food and farming research agenda 2025. Organic knowledge for the future. *Technology Platform Organics*, http://orgprints.org/13439/

Olesen, J. E., Schelde, K., Weiske, A., Weisbjerg, M. R., Asman, W. A. H., & Djurhuus, J. (2006). Modelling greenhouse gas emissions from European conventional and organic dairy farms. *Agriculture, Ecosystems & Environment, 112*, 207–220.

Pacini, C., Wossink, A., Giesen, G., Vazzana, C., & Huirne, R. (2003). Evaluation of sustainability of organic, integrated and conventional farming systems: A farm and field-scale analysis. *Agriculture, Ecosystems & Environment, 95*, 273–288.

Pacini, C., Giesen, G., Wossink, A., Omodei-Zorini, L., & Huirne, R. (2004). The EU's Agenda 2000 reform and the sustainability of organic farming in Tuscany: ecological-economic modelling at field and farm level. *Agricultural Systems, 80*, 171–197.

Pavie, J., & Lafeuille, O. (2009). *Valorisation des données technico-économiques des exploitations allaitantes biologiques suivies dans les réseaux d'élevage. Evolutions d'un échantillon constant sur 7 campagnes (2000–2006).* Réseaux d'élevage pour le conseil et la prospective, Collection Références, Institut de l'Elevage et Chambres d'Agriculture.

Pelletier, N., Pirog, R., & Rasmussen, R. (2010). Comparative life cycle environmental impacts of three beef production strategies in the Upper Midwestern United States. *Agricultural Systems, 103*, 380–389.

Phetteplace, H. W., Johnson, D. E., & Seidl, A. F. (2001). Greenhouse gas emissions from simulated beef and dairy livestock systems in the United States. *Nutrient Cycling in Agroecosystems, 60*, 99–102.

Refsgaard, K., Halberg, N., & Kristensen, E. S. (1998). Energy utilisation in crop and dairy production in organic and conventional livestock production systems. *Agricultural Systems, 57*(4), 599–630.

Réseau d'Elevage Charolais. (2006). *Bassin Charolais. Conjoncture économique des systèmes bovin viande, campagne 2006.* Document, Chambre d'Agriculture, Institut de l'Elevage, 50 p.

Siciliano, G. (2009). Social multicriteria evaluation of farming practices in the presence of soil degradation: A case study in southern Tuscany, Italy. *Environment, Development and Sustainability, 11*(6), 1107–1133.

Simon, J. C., & Le Corre, L. (1992). Le bilan apparent de l'azote à l'échelle de l'exploitation agricole: méthodologie, exemples de résultats. *Fourrages, 129*, 79–94.

Soussana, J. F., Fuhrer, J., Jones, M., & Van Amstel, A. (2007). The greenhouse gas balance of grasslands in Europe. *Agriculture, Ecosystems & Environment, 121*, 1–4.

Vergé, X. P. C., Dyer, J. A., Desjardin, R. L., & Worth, D. (2008). Greenhouse gas emissions from the Canadian beef industry. *Agricultural Systems, 98*, 126–134.

Veysset, P. (2002). *Conversion of mixed holdings (crop/suckler cattle) to organic farming: Finding a balance between fodder and cash crops and economic consequences.* (In: Proceedings of the 14th Organic World Congress, IFOAM, 117.)

Veysset, P., Bébin, D., & Lherm, M. (2005). Adaptation to agenda 2000 (CAP reform) and optimization of the farming system of French suckler cattle farms in the charolais area: A model-based study. *Agricultural Systems, 83*, 179–202.

Veysset, P., Bécherel, F., & Bébin, D. (2009). Elevage biologique de bovins allaitants dans le Massif Central: résultats technico-économiques et identifications des principaux verrous. *Inra Productions Animales, 22*(3), 189–196.

Veysset, P., Lherm, M., & Bébin, D. (2010). Energy consumption, greenhouse gas emissions and economic performance assessments in French Charolais suckler cattle farms: Model-based analysis and forecast. *Agricultural Systems, 103*, 41–50.

Wood, R., Lenzen, M., Dey, C., & Lundie, S. (2006). A comparative study of some environmental impacts of conventional and organic farming in Australia. *Agricultural Systems, 89*, 324–348.

Chapter 11
Assessing Multifunctionality in Relation to Resource Use: A Holistic Approach to Measure Efficiency, Developed by Participatory Research

Johanna Björklund and Börje Johansson

Abstract In this study, emergy analysis and footprinting were combined to assess and illustrate the total resource use caused by milk production and to identify the renewable fraction of this resource use. The total efficiency was defined as a function of the resource use and the multifunctionality of production. The classification of ecosystem services in the Millennium Ecosystem Assessment (MA) was used as the basis for ranking multifunctionality. Three scenarios with different degrees of input intensity and milk production were constructed and compared with the current production mode. The ratio of local renewable resource use to total resource use differed greatly between the different production strategies, being 1:3 for a self-sufficient organic farm and 1:14 for a conventional farm with maximum milk yield. Milk production was fivefold higher on the conventional farm, while generation of ecosystem services increased with increasing self-sufficiency under the local conditions prevailing in the study. Ecosystem services in all categories except provisioning were ranked higher when self-sufficiency increased.

Keywords Emergy analysis • Footprint • Ecosystem services • Ecosystem bundles

---

J. Björklund, Ph.D. (✉)
Man-Technology-Environment, School of Science and Technology, Örebro University, 701 82 Örebro, Sweden
e-mail: johanna.bjorklund@oru.se

B. Johansson
Farmer, Hulta Norrgård, 58596 Linköping, Sweden
e-mail: hulta.norrgard@privat.utfors.se

A.A. Marta-Costa and E. Silva (eds.), *Methods and Procedures for Building Sustainable Farming Systems: Application in the European Context*,
DOI 10.1007/978-94-007-5003-6_11, © Springer Science+Business Media Dordrecht 2013

11.1 Introduction

The global environmental crisis clearly shows the need for assessment methods that are able to relate to the complexity of the natural world and help us transform the current food system into a sustainable system (Rockström et al. 2009). We need to develop a system that can sustain production of sufficient food, allowing fair distribution and maintaining production capacity in the long run. This requires a production system that does not compromise the health and integrity of local and global ecosystems.

Global food demand is projected to increase by up to 70% in the coming 50 years (FAO 2009). This challenge will be enormous, taking into account that large areas formerly known as the world's granaries are likely to be compromised by the global environmental crisis (Brown 2009; IPCC 2007).

Furthermore, the United Nations-initiated Millennium Ecosystem Assessment (MA) found that about 60% of the assessed ecosystem services were being used faster than their rate of regeneration. Fundamental services for human well-being, and not least for food production, are in danger (Millennium Ecosystem Assessment 2005). This concerns services such as the availability of fresh water, the productive capacity of seas, the maintenance of genetic resources and the ability to mitigate natural hazards. Human activities have now exceeded at least three of the nine boundaries within which we need to remain in order to function safely on Earth. With our present impact on the global climate and on global nitrogen cycling, as well as our contribution to the rapid reduction in biodiversity, we can expect unpredictable and uncontrollable changes (Rockström et al. 2009). Added to this is the issue of peak production of energy from fossil fuels (Bentley 2002). This will have dramatic effects on the industrial mode of food production predominant in many countries today. Abundant access to cheap energy has been fundamental for an agricultural system that unreservedly relies on fossil-fuel-based inputs.

The development of farming systems that have high water-, nutrient- and energy-use efficiencies and conserve natural resources and biodiversity without compromising yield is of the utmost importance. International research and consultation bodies such as FAO (Food and Agriculture Organization of the United Nations) and IAASTD (International Assessment of Agricultural Science and Technology for Development) have concluded that larger opportunities lie in small-scale farming than previously thought (Beintema et al. 2008; FAO 2007; OngÅLwen and Wright 2007). One important reason identified was the high productivity and the generation of environmental services in such agriculture.

When assessing the multifunctionality of an agricultural activity, multi-criteria tools to evaluate productivity and efficiency need to be employed. Existing methods, which measure one aspect at a time and focus either on monetary values or on biophysical issues, are unsuitable (Cuadra and Björklund 2007). Ecological values, e.g. generation of ecosystem services, are often not considered at all due to lack of reliable and comparable measurements (Björklund and Rydberg 2003).

There is currently a lack of assessment methods that relate resource use to the total outcome of an agricultural activity, that evaluate the efficiency of the production (including food products) as well the generation of cultural and natural services and values and that distinguish between local/purchased and renewable/non-renewable resources. Such assessment methods must have system boundaries wide enough in time and space to provide a basis for constructive dialogue and informed decisions on changes.

The present study combined existing assessment methods such as emergy analysis (Odum 1996) and ecological footprinting (Wackernagel and Rees 1996). Emergy analysis accounts for the energy intensity embodied in products, including the environmental work and the work of humans in generating products and services, while ecological footprinting presents the resource use on an area basis. Combining these methods can allow the total resource use on a farm and the part that is local and renewable to be determined.

The classification of ecosystem services in the Millennium Ecosystem Assessment (MA) was used here as the basis for ranking multifunctionality, and the results were expressed in the form of ecosystem bundles (see, e.g. Foley et al. 2005). Assessing the multifunctionality of the farm produced a more perceptive measurement of the efficiency of resource use than if we had assigned the resource use only to the production of, e.g. milk or meat.

The overall aim of the study was to develop a useful tool to assess whether changes implemented at farm level to reduce the dependence on fossil-fuel-based inputs and the associated impact on global warming actually do so in a longer and broader perspective. The study compared the current production mode on an organic dairy farm with scenarios of alternative production modes with different degrees of intensity in external inputs and milk yields. Specific objectives were to obtain knowledge on options to reduce fossil fuel use and the impact on global warming and to contribute to the development of multi-criteria assessment tools.

The multi-criteria method and the research question: *"How do we know that what we are doing is actually leading to more efficient resource use and is employing more local ecosystem services?"* were formulated by a participatory research group in Sweden comprising farmers and researchers.

11.2 Materials and Methods

The study was performed on an organic dairy farm (A), with three scenarios for alternative production strategies: organic milk production maximising self-sufficiency in feed and energy (B), organic production maximising milk yields (C) and conventional production maximising milk yields (D) (Table 11.1).

The study was carried out in 2007–2009 as part of a participatory research project involving 11 farmers on eight certified organic farms in central Sweden, a researcher

Table 11.1 Some data on current production of the organic dairy farm studied (A) and the three alternative scenarios (B–D), all comprising different modes of dairy production

	(A) Current production	(B) Low-input organic production	(C) Organic production, high yield	(D) Conventional production, high yield
Agricultural land (ha)	70	70	70	70
Arable land (ha)	40	40	30	30
Ley on arable land	30	30	30	30
Managed natural grazing (ha)	30	30	10	0
Fallow	0	0	30	40
Average field size (ha)	0.5	0.5	2	2
Milking cows (number)	18	18	40	64
Animal units (number/ha)	0.88	0.88	1.33	1.6
Sheep (number)	25	25	0	0
Milk production (kg cow/year)	7,000	6,000	9,000	10,000

from the Swedish University of Agricultural Sciences and a process facilitator. The project sought to identify and communicate ways to develop sustainable and fair farming systems based on local ecosystem resources and services. Important aspects of the work were the joint learning process and helping to inspire and open up opportunities to talk about agricultural sustainability with other farmers, consumers and decision-makers. On the basis of research results combined with their practical experiences, the group of farmers helped develop useful tools and methods for assessment.

11.2.1 Description of the Farm and the Three Production Scenarios

11.2.1.1 The Study Farm and Its Existing Production System

The farm Hulta Norrgård is situated in a mainly forested small-scale mosaic area at the edge of agricultural plains, about 20 km south of the city of Linköping in south-east Sweden. The farm is organically certified, with dairy production as the main enterprise. The farm comprises 70 ha of arable land and grazing, and field size is on average 0.5 ha. The animal herd consists of 18 dairy cows, 25 sheep, a few hens and ducks, one dog used to herd the animals out to pasture and some cats. The animals provide the basis for nutrient recycling of the farm, and the majority (97%) of the animal feed is produced on the farm (Table 11.1). The farm also includes 30 ha of forest that were not included in the study although a few hectares of this forest are sometimes used as pasture for cows. The milk production is around 7,000 kg per cow and year. The farmer and his wife work full-time on the farm.

The production strategy on the farm is to minimise external resource use so as to become resource-efficient, keep costs low and achieve adequate profitability. Some feeds, seeds and diesel are brought into the farm. The machinery is old, and if it breaks down, it is repaired on the farm.

The farm is a node in the local recycling association, and it collects and uses the urine and faeces from half of the approximately 45 households in the three villages that comprise the association. The association, which has been in operation for 14 years, takes local recycling and local food as the starting point for rural development. In a small shop on the farm, villagers can sell and buy local products, such as eggs, milk, honey and furs, as well as some fair-traded imported products. Among other things, the farm also provides and prepares land for common cultivation of potatoes and organises hay-making on old meadow in an annual festival.

11.2.1.2 Scenarios for Alternative Production Modes

Organic Dairy Production Maximising Self-Sufficiency in Feed and Energy

In this scenario, all animal feed is produced on the farm. The feed concentrate used is rapeseed cake produced on the farm. The rapeseed oil is used as fuel for the tractors. The cows are mainly fed on grass-clover silage and hay. The size of the dairy herd remains the same (18 cows), but milk production has declined to 6,000 kg per cow and year. Sheep, hens and ducks are kept as today. All land is used, arable land as well as natural pastures and meadows. The sale of local products, e.g. milk and eggs, has been increased and contributes to the farm economy. Wind electricity is produced on-farm to meet the energy demands of the farm.

The strategy is to minimise purchased inputs as much as possible. Food for the family is also produced on the farm: Cereals for own consumption are processed in a small mill, vegetables are grown, and eggs and meat are produced. There is as little machinery as possible, and the machines are old and kept repaired on the farm. Two people work full-time on the farm, and the cash withdrawals are low.

Local recycling is strengthened, and crop seeds are produced on-farm and exchanged with neighbours.

Organic Production Maximising Milk Yields

The dairy herd is more than doubled, to 40 cows. The majority of the fodder is purchased in, including all feed concentrate and a large proportion of the grain. The most distant pastures and the smallest fields are abandoned. Sheep and poultry are not kept. Dairy production is optimised to give as high milk yields as possible while still conforming to the rules for organic certification. Silage and hay comprise 60% (by weight) of the feed on an annual basis. Animals are mainly fed indoors year around, with only a small proportion of their diet coming from pastures near the animal house. Milk production is around 9,000 kg per cow and year.

The strategy is to increase the profitability of the farm, both by having high milk production and by adding value to the production by being organically certified. Machinery is modern and investments are made in labour-saving techniques. One person works full-time; hired labour is used for relief milking.

There is no selling of local products, but local recycling of nutrients from households to the farm continues.

Conventional Production Maximising Milk Yields

The dairy herd is increased to 64 cows, which is the maximum amount in terms of stocking density regulations on the area required for manure application. Only milking cows are kept on the farm, with recruitment leased out to a farm specialising in dairy heifer production. Animals are fed indoors during milking all year around. Only 10 ha of land near the animal house are used to meet Swedish regulations on the period of outdoor grazing provided for dairy cows. No natural pastures are maintained as the grass contains inadequate concentrations of nutrients to act as feed for high-milking cows. All feeds except grass silage and hay are bought. Fertiliser, herbicides, pesticides and fungicides are used regularly. Milk production is around 10,000 kg per cow and year.

The strategy is to increase the profits from the farm, e.g. by decreasing labour costs and increasing income. One person works full-time on the farm. Hired labour is used for relief milking at weekends on a regular basis. The machinery is modern and the animal house and milking parlour are easy to work in.

There is no local recycling of nutrients as there is no perceived economic, ecological or social incentive for this, and it would only increase the workload.

11.3 Emergy-Based Footprint

Emergy analysis accounts for the energy intensity embodied in products, including the environmental work and the work of humans in generating products and services. All items are converted to a common basis of solar energy (the unit solar emjoules (sej)) using conversion coefficients (transformities). The transformity is the solar energy used to make one joule of a resource (Odum 1996). The method has been comprehensively described by a number of authors (Björklund 2000; Brown and Herendeen 1996; Odum 1996).

Emergy analysis has been used in this study to assess and compare resource of different kinds, such as the use of fossil fuels, of iron and of energy in rain. With the emergy analysis, it has been possible to put all different resources on a common basis, of the amount of work of environment (the solar energy used) to generate them, and make a comparison that is consistent. By using emergy analysis, it was possible to include all resources in the same analysis.

To account for emergy in purchased goods, both the emergy input from the environment to generate the raw material and the emergy in human services to make the raw material useful in the economic system were calculated. Emergy in service was calculated from the average emergy flow per unit of money for Sweden (Lagerberg et al. 1999). Emergy caused by resource-use related to farmers' own labour was omitted in all calculations due to difficulties in relating economic withdrawal and farm labour.

To facilitate informed discussion, the emergy use was converted to area by dividing it by annual renewable emergy inflow per m^2 (approximately 5.70×10^{10} sej m^{-2} year^{-2}, which was the contribution from rain). In this way, an indirect area demand was calculated for all purchased inputs. This allowed comparison of the direct and indirect area demand for the different systems. We called this approach the "emergy-based footprint", and it corresponded to the theoretical area needed if all resources used on the farm were local, renewable resources. This has the same conceptual basis as the ecological footprint developed by Wackernagel and Rees, but the system boundaries are larger and the calculations are based on energy flows, while the Wackernagel and Rees method calculates biological production (Odum 1996; Wackernagel and Rees 1996).

11.4 Bundles of Ecosystem Services

Multifunctionality was assessed as the kind and amount of generation of ecosystem services, adopting the classification of ecosystem services used in the Millennium Ecosystem Assessment (MA) (Millennium Ecosystem Assessment 2005). Ecosystem services are absolutely vital for human civilisations and can be defined as "... conditions and processes through which natural ecosystems, and the species that make them up, sustain and fulfill human life" (Daily 1997).

A ranking tool was developed to assess the generation of the services, and the results were described in the form of bundles of ecosystem services (see, e.g. Foley et al. 2005, The Resilience Alliance (www.resalliance.org/3683.php, visited 15 January 2010)). Ten people, including the farm family (regarded as one individual), two agricultural researchers and seven farmers, were asked to use short descriptions to rank their opinion of the potential for generation of ecosystem services on a scale from 1 (the lowest ranking) to 5 (the highest) for the farm and the three alternative production scenarios. Weighted averages were then calculated for each service. Eight services, two from each of the four main classifications of MA (provisioning, supporting, regulating, cultural), were depicted in the bundle. The two chosen were important services in relation to agriculture while also being perceived as representative for that class. Each individual ecosystem service was named according to the MA system as far as possible, but due to the focus of the study, names were modified or services subdivided to be more specific when appropriate. However, the ranking for the provisioning service (milk production) was assessed by the actual milk yield estimated in the scenario, based on diet. The perceived greatest potential of an agroecosystem to generate a specific ecosystem service was used as a reference point for the ranking.

11.5 Results

11.5.1 Assessment of Resource Use

The emergy-based footprint, describing the total area needed when all resources used in milk production were produced locally, ranged from 3 times the actual farm area (option B, organic high self-sufficiency) to 14 times that area (C, conventional high yielding) (Fig. 11.1). In the current production system (A), the total footprint

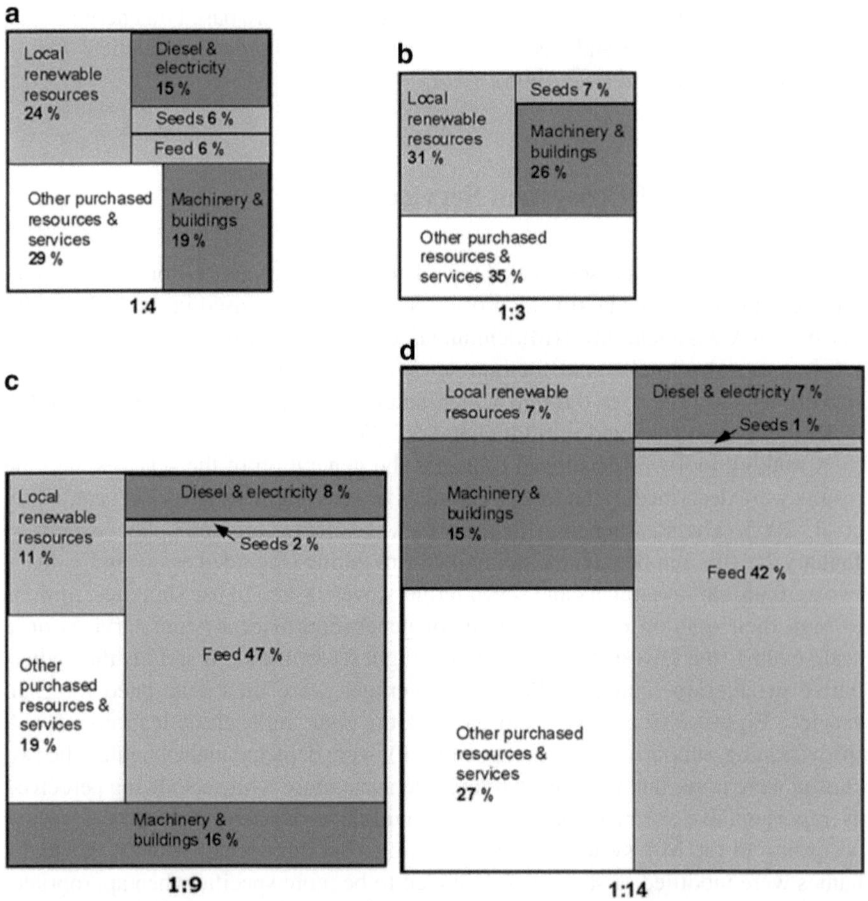

Fig. 11.1 Emergy-based footprint for the current production system (**a**) and for three alternative production scenarios: (**b**) low-input organic production, (**c**) organic production maximising yield and (**d**) conventional production maximising yield. The ratios below each picture show the local renewable resource use, which is the actual agricultural land area comprising the farm, to total resource use for the production. The agricultural land is similar for all production alternatives (local renewable resources) but differs in proportion to all other resource use

was 4 times the actual farm area. The local resource use was considered renewable as no non-renewable resources such as peat were used, soil organic carbon was maintained at a stable level, and there were no signs of soil erosion. Not surprisingly, the local renewable resource use decreased with decreasing self-sufficiency (31% in option B, which maximised self-sufficiency; 24% in the current system A; 11% in option C, which maximised organic milk yield; and 7% in option D, which maximised conventional milk yield).

Purchased feed accounted for the largest resource use in the scenarios where milk production was maximised (47% in the organic option (C) and 42% in the conventional (D)). In the scenarios with a high degree of self-sufficiency, the greatest resource use was purchased services (other resource use covered by money flows), e.g. veterinary costs, interest and insurances (29% in current production (A) and 36% in the option maximising self-sufficiency (B)).

The proportion of resources from machinery and buildings decreased as milk production per cow increased (26% in the option maximising self-sufficiency (B), 19% in current production (A), 16% in the system maximising organic milk yield (C) and 15% in the option maximising conventional milk yield (D)).

Lagerberg et al. (1999) estimated the average renewable part of the resources driving the Swedish economy to be 13% of total resources used. Using that figure, the proportion of the resource use in our scenarios that could be considered renewable (although not only local, as it also included purchased resources) ranged from 35% in current production (A) to 50% for the scenario maximising organic milk yield (C). For the conventional scenario (D), the renewable part was 37%, while for the scenario maximising self-sufficiency (B), it was 40%.

11.5.2 Assessment of the Output of the Farm: The Multifunctionality

The illustration of the multifunctionality obtained by ranking the generation of ecosystem services in the different scenarios indicated that high milk production conflicts with the generation of other ecosystem services.

The current production system received a high ranking in all MA classes, ranging from 4.3 for cultural services to 3.2 for provisioning services (Fig. 11.2a). Nutrient recycling and soil fertility were the individual services that scored highest, 4.4 and 4.3, respectively, while milk production and contribution to climate regulation got the lowest scores, 1.0 and 3.6, respectively. The picture was similar for the scenario maximising self-sufficiency (B), with a ranking of 4.8 for nutrient recycling and only 0.8 for milk production (Fig. 11.2b).

For the scenarios maximising milk yields, the situation was reversed, with high scores for milk production, 5.0 for conventional production (D) and 2.8 for organic (C) (Fig. 11.2c, d). For the organic option, however, soil fertility got the highest ranking, 3.6. The contribution to the generation of genetic resources, wild and domestic, was ranked lowest (1.2) for the conventional production option, while the contribution to climate regulation was lowest (2.0) for the organic production option.

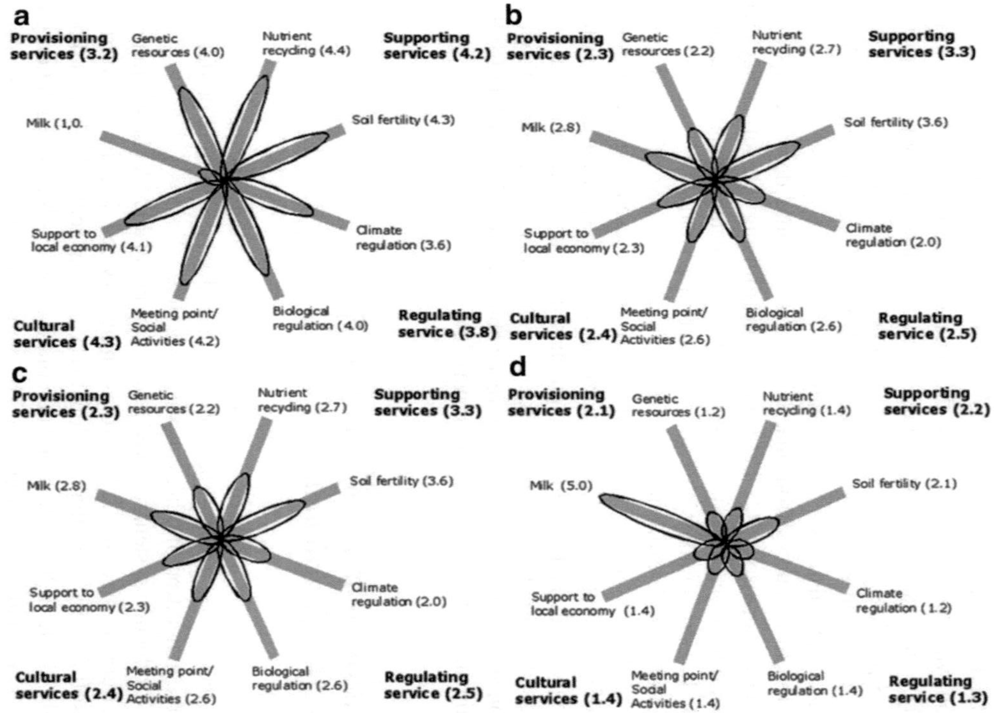

Fig. 11.2 Assessment of generation of ecosystem services for the current production system (**a**) and for three alternative production scenarios: (**b**) low-input organic production, (**c**) organic production maximising yield and (**d**) conventional production maximising yield. Classification system in the Millennium Ecosystem Assessment was used as basis for the assessment, which was done as a ranking exercise by farmers and researchers individually on basis of mainly written information about the farm and the scenarios constructed. Figures in parenthesis are weighted averages of rankings of certain services as well as of all the services in the different classes

When the average scores for all four classes were added up for the four scenarios, that maximising self-sufficiency (B) received the highest score of 16.5, the current production system (A) scored 15.5, the system maximising organic milk production (C) scored 10.5 and that maximising milk conventional production scored 7.0.

11.6 Discussion

In the present study, the degree of self-sufficiency seemed to have a larger impact on the size of the emergy-based footprint than whether the production was organic or conventional. Purchased feed was the item making the largest contribution to the footprint when milk production was maximised.

The total size of the footprint did not differ substantially when the production was organic when the strategy was to maximise milk production (Fig. 11.1c, d). However, the share of the resource use that was estimated to be renewable, local and not local (results not shown) was higher when production was organic. In the scenario where organic milk production was maximised, this share was the highest for all four scenarios. The reason was a large proportion of renewable resources in purchased feed.

Even when the multifunctionality of the production was considered and expressed in ecosystem bundles, the differences were larger between the scenarios with high and low self-sufficiency than between organic and conventional.

The high ranking of nutrient recycling for the scenarios with high self-sufficiency (Fig. 11.1a, b) was obviously due to no or nearly no purchased feed and recycling of sewage from neighbouring households. Local selling of milk was also highest in option B.

A reason why the scenario maximising organic milk production (C) scored higher than that maximising conventional milk production (D) for soil fertility, even though the area of ley production was similar, may be the common use of a larger number of different sorts and varieties of ley species including clover (Trifolium) in organic agriculture than in conventional. Fields with high biodiversity have been shown in field trials to have a higher potential to build-in soil carbon than fields with lower biodiversity, even when yields are similar (Steinbeiss et al. 2008). This may refer also to the capacity to contribute to climate regulation, which accordingly was ranked higher for the organic scenarios.

A conceivable argument for the higher ranking of the potential for maintenance of genetic resources, wild and domestic, and for biological regulation for the scenarios with high self-sufficiency and lower production intensity is that the habitat variety increases when the area is maintained with lesser intensity. Moreover, the habitat diversity is larger, e.g. when a cropping sequence includes cereals, legumes, oilseeds and ley than only one of these crops, as in the scenarios maximising milk production (leys in this study). High habitat diversity is suggested to be one of the main factors producing potential for high biological diversity in agriculture (Benton et al. 2003). In contrast, high intensity in production (measured as harvested yields) has been found to reduce the biodiversity (Donald et al. 2006). An obvious reason for the low score for strongly biodiversity-related services in conventional production is the use of herbicides and pesticides. Moreover, old breeds of animals and varieties of crops are also maintained to a higher degree in alternatives with a high degree of self-sufficiency, as they have been shown to be more resilient and yield better when soil nutritional levels are low and variable (Araya and Edwards 2006; OngÅLwen and Wright 2007).

Selling produce locally and the involvement in the local recycling association contributed to local economy and at the same time provided more space and motivation for social activities in relation to the farm in the scenarios with the strategy of high self-sufficiency compared with the scenarios with a strategy to maximise milk production.

When the average ranking of ecosystem services for all classes was calculated, the scenarios with high self-sufficiency achieved the highest values. This may indicate that they are the most multifunctional production systems. However, the final value masks the fact that the ranking is relative and that some services may be more vital than others in a global or local sense and during different times. A high score in that case would depend on the ability of other areas to provide the specific service and also what is most urgent to consider at that time. Two obvious examples are milk production if people are starving locally or globally and climate regulation when emissions are too high in society in general. Changing perspective from an anthropocentric to an ecocentric focus may also shift the value of the different services.

The perceptions of persons performing the rankings, which were based on their knowledge of the systems and their former experiences, as well as what they consider to be the reference points, had a large impact of the results. Assessments by a large group of people from different fields would have made the ranking more reliable. The choice of individual ecosystem services or functions presented from the ranking was also crucial for the results presented.

Finally, the construction of the scenarios and the estimates of resource use made for calculation of the emergy-based footprint were found to be crucial for the results. A sensitivity analysis would have made the results more reliable. In spite of this, the participatory research group found that this multi-criteria method provided information to deepen the discussion on what a future sustainable agricultural system would look like. The method was perceived as well worth further development and application in other assessments.

Acknowledgements Thanks to Ebba and Sven Schwartz foundation for funding this study.

References

Araya, H., & Edwards, S. (2006). The Tigray experience. A success story in sustainable agriculture. *Third World Network, Environment & Development Series 4,* Penang, Malaysia, p. 54

Beintema, N., Bossio, D., Fernandez, M., Gurib-Fakim, A., Hurni, H., Izac, A. -M., Jiggins, J., Kranjac-Berisavljevic, G., Leakey, R., Ochola, W., Osman-Elasha, B., Plencovich, C., Roling, N.,

Rosegrant, M., Rosenthal, E., Smith, L. (2008). *International Assessment of Agricultural Knowledge, Science and Technology for Development (IAASTD) Global Summary for Decision Makers*. www.agassessment.org

Bentley, R. W. (2002). Global oil & gas depletion: An overview. *Energy Policy, 30*, 189–205.

Benton, T. G., Vickery, J. A., & Wilson, J. D. (2003). Farmland biodiversity: Is habitat heterogeneity the key? *Trends in Ecology & Evolution, 18*(4), 182–188.

Björklund, J. (2000). *Emergy analysis to assess ecological sustainability, strengths and weaknesses*. Dissertation, Agraria 242, SLU, Uppsala.

Björklund, J., & Rydberg, T. (2003). *Att värdera uthållighet i lantbrukets markanvändning – genomgång av metoder för miljö- och naturresursanalys*. Centrum för uthålligt lantbruk, SLU, Uppsala (In Swedish).

Brown, L. R. (2009). *Plan B 4.0 Mobilizing to save civilization*. New York: Earth Policy Institute.

Brown, M. T., & Herendeen, R. (1996). Embodied energy analysis and EMERGY analysis: A comparative view. *Ecological Economics, 19*, 219–236.

Cuadra, M., & Björklund, J. (2007). Assessment of economic and ecological carrying capacity of agricultural crops in Nicaragua. *Ecological Indicators, 7*(1), 133–149.

Daily, G. C. (Ed.). (1997). *Nature's services. Social dependence on natural ecosystem services*. Washington, DC: Island press.

Donald, P. F., Sanderson, F. J., Burfield, J. I., & van Bommel, F. P. J. (2006). Further evidence of continent-wide impacts of agricultural intensification on European farmland birds, 1990–2000. *Agriculture, Ecosystems & Environment, 116*, 189–196.

FAO. (2007). *The state of food and agriculture. Paying farmers for environmental services*. http://www.fao.org/docrep/010/a1200e/a1200e00.htm

FAO. (2009). *How to feed the world in 2050. Discussion paper*. Rome, 12–13 October 2009. http://www.fao.org/wsfs/forum2050/wsfs-background-documents/issues-briefs/en/

Foley, J. A., DeFries, R., Asner, G. P., Barford, C., Bonan, G., Carpenter, S. R., Chapin, S. F., Coe, M. T., Daily, G. C., Gibbs, H. K., Helkowski, J. H., Holloway, T., Howard, E. A., Kucharik, J., Monfreda, C., Patz, A., Prentice, C., Ramankutty, N., & Snyder, P. K. (2005). Global consequences of land use. *Science, 309*, 570–574.

IPCC. (2007). *Impacts, adaptation and vulnerability. Working group II. Contribution to the Intergovernmental Panel och Climate Change Fourth Assessment Report. Summary for Policymakers*. www.ipcc.ch

Lagerberg, C., Doherty, S. J., & Nilsson, P. O. (1999). Evaluation of the resource use efficiency and sustainability of the Swedish economy using emergy based indices. In C. Lagerberg (Ed.), *Emergy analysis of the resources use in Greenhouse Crop Production and of the resources basis of the Swedish economy*. Sweden: Swedish University of Agricultural Sciences.

Millennium Ecosystem Assessment. (2005). *Ecosystems and human well-being: Synthesis*. Washington, DC: Island Press.

Odum, H. T. (1996). *Environmental accounting: Emergy and environmental decision making*. New York: Wiley.

OngÅLwen, O., & Wright, S. (2007). *Small farmers and the future of sustainable agriculture*. Berlin: Heirich Böll Foundation, Misereor, Wuppertal Institute, Wuppertal.

Rockström, J., Steffen, W., Noone, K., Chapin, F. S., III, Lambin, E., Lenton, T. M., Scheffer, M., Folke, C., Schellnhuber, H. J., Nykvist, B., De Wit, C. A., Hughes, T., Van der Leeuw, S., Rodhe, H., Sörlin, S., Snyder, P. K., Costanza, R., Svedin, U., Falkenmark, M., Karlberg, L., Corell, R. W., Fabry, V. J., Hansen, J., Walker, B., Liverman, D., Richardson, K., Crutzen, P., & Foley, J. (2009). Planetary boundaries: Exploring the safe operating space for humanity. *Ecology and Society, 14*(2), 32 [Online].

Steinbeiss, S., Beßler, H., Engels, C., Temperton, V. M., Buchmann, N., Roscher, C., Kreutziger, Y., Baade, J., Habekost, M., & Gleixner, G. (2008). Plant diversity positively affect short-term soil carbon storage in experimental grasslands. *Global Change Biology, 14*, 2937–2949.

Wackernagel, M., & Rees, W. (1996). *Our ecological footprint. Reducing human impact on the earth*. Gabriola Island: New Society Publisher.

Chapter 12
Indicator-Based Method for Assessing Organic Farming Sustainability

Mohamed Gafsi and Jean Luc Favreau

Abstract Given the crisis of productivism, organic agriculture has in recent years a remarkable development. It appears in fact as a form of sustainable agriculture alternative to conventional agriculture. But despite this strong development and the growing importance, there is a lack of methods for assessing the sustainability of organic farms. There was an abundant production of methods for assessing the sustainability of farms; but most often, these methods are intended for conventional farming systems and therefore take little account of the specificity of organic farming. In this chapter, we propose a method for assessing farm's sustainability suitable for organic farming. This method relies on one side of the agricultural sustainability principles and the other side on the principles and characteristics of organic farming. The results of application of this method on a sample of farms in the Midi-Pyrenees region (southern France) show that it takes into account very well the agro-ecological and socio-territorial specificities of organic farming.

Keywords Organic farming sustainability • Indicator-based method • Systemic approach • Agro-ecological sustainability • Socio-territorial sustainability

12.1 Introduction

Agricultural researchers and professionals widely recognise the importance of sustainable agriculture and the need to make this operational, i.e. to develop appropriate methods to measure sustainability of farming system (Webster 1997; Kropff et al. 2001; Van Calker et al. 2006; Gafsi et al. 2006). There is an abundant scientific

M. Gafsi (✉) • J.L. Favreau
National School of Agronomic Training, UMR Dynamiques Rurales,
1 Route de Narbonne, 31326 Castanet-Tolosan, France
e-mail: mohamed.gafsi@educagri.fr; jeanluc.favreau@free.fr

A.A. Marta-Costa and E. Silva (eds.), *Methods and Procedures for Building Sustainable Farming Systems: Application in the European Context*,
DOI 10.1007/978-94-007-5003-6_12,

literature on the topic of sustainability assessment (Zander and Kachele 1999; Andreoli and Tellarini 2000; Rigby et al. 2001; Heller and Keoleian 2003; Ness et al. 2007; Sydorovuch and Wossink 2008). Consequently, an increasing variety of evaluation methods for assessing the sustainability, notably at the farm's level, have been produced (Van der Werf and Petit 2002; Bockstaller et al. 2009).

But most often, these methods are intended for conventional farming systems and therefore take little account of the specificity of organic farming. While it is conventional agriculture which raises more concerns over the adverse effects of cropping and farming systems such as water pollution by nitrates and pesticides and gaseous emissions due to nitrogen inputs. But sustainability is not just a matter of treating adverse effects of productivism. Other factors outside the conventional system can lead to lack of sustainability. Despite not having the negative effects of productivism, organic farming can present lack of sustainability. The sustainability of organic farming systems needs therefore to be appreciated. However, in fact, the sustainability of organic farming is addressed in the context of comparison between organic, integrated and conventional farming systems, using the same sets of indicators (Vereijken 1997; Pacini et al. 2003). In France, there are currently a dozen indicator-based methods used by professional actors and teaching, but none are specific to organic farming.

This contrasts with the emphasis today in organic farming and its recent dynamics of development. Over the past 15 years, organic farming has been in France a very strong development both at the production level and at the demand of organic products. In 1995, 3,600 organic farms were farming roughly 130,000 ha; in 2008, it was 13,298 farms on 583,799 ha (2,12% of the total national usable agricultural area). In the meantime, food processing and marketing companies using an organic label had grown from 700 to 7,398 (Agence Bio 2009). The gross market of organic products has the same dynamics; the average annual growth is more than 10% since 1999, whence the global food market was growing by 3% yearly. Besides, as evidenced by the decisions of the Grenelle de l'Environnement forum in 2007 and the French national plan for the development of organic farming ("Organic farming: Towards 2012", Ministry of Agriculture[1]), organic farming is now seen as a sustainable alternative to conventional agriculture. But despite this general context of development, organic farmers demonstrate several technical, economic, organisational and other difficulties. This raises the following questions: Are these farms sustainable? How to assess their sustainability?

In this chapter, we propose a method for assessing farm's sustainability suitable for organic farming. This method relies on one side of the indicator-based methods of farms' sustainability and the other side on the principles and characteristics of organic farming. It is based on a comprehensive and systemic approach to sustainable farming. The concept of sustainability is presented in three dimensions (economic, environmental and social); each one includes some components that are,

[1] The plan aims to increase of surfaces to reach in 2012, 6% of cultivated areas, to provide aids for conversion, farmers' training, etc.

themselves, divided in several indicators. The three dimensions of sustainability are now well known, and they are the subject of a consensus between different actors both scientific and professional. Our methodological work focuses so on the levels of components and indicators. It consists of four steps: (i) determination of list of components, for each dimension, that reveal the specificity of organic farming systems; (ii) identification and selection of set of indicators for each component; (iii) estimation of relative weights of different components in the overall sustainability measure, which is crucial in the aggregation process; and (iv) setting thresholds and scoring system for each indicator. To illustrate possible applications of the proposed method, we studied the sustainability of selected farms in the region of Midi-Pyrenees (southern France). These farms are chosen regarding their dimension and their production systems orientation. Our objective is mainly to demonstrate how the proposed method could be used and to assess its performances.

This chapter proceeds as follows: the theoretical framework for our analysis is presented in Sect. 12.2, where we review the relevant literature on agricultural sustainability assessment and organic farming principles. Section 12.3 discusses the specifics of the research design including the method design and the fieldwork presentation. In Sect. 12.4, we present the results of the application of the proposed method to a sample of organic farms. We conclude in Sect. 12.5 with the discussion of our findings regarding the method, its application in the case study and its applicability in other contexts.

12.2 Theoretical Framework of Sustainability Assessment of Organic Farming Systems

12.2.1 Sustainability and Sustainability Measures

The need for definition of sustainable agriculture is a prerequisite for developing an assessment framework. Although there is no single definition of sustainability, there are major common features that are defined (e.g. Hansen and Jones 1996; Park and Seaton 1996; Rigby et al. 2001; Godard and Hubert 2002). Sustainable agriculture should be the ability of farming systems to maintain its productivity and usefulness to society in the long term. This means that sustainable agriculture includes both the long-term viability of farming system itself and the contribution of this farming system to the sustainability of the territory and the communities to which it belongs. The second aspect in this definition is crucial for the meaning of sustainable agriculture, and it must be considered in our assessment framework of sustainability of organic farming systems. It places farmers squarely within the local social fabric, offering local services, maintaining and creating jobs in the rural space, contributing to rural planning, developing environmental services, dealing with negative external effects on the environment, etc. To take care of those two aspects (i.e. viability of farming system and its contribution to the sustainability of the territory), a sustainable farming system must at the same time be economically viable, ecologically sound and socially responsible (Ikerd 1997). Consequently, an assessment framework of sustainability requires an integrated and holistic approach which is addressing in the same time different and competing objectives (Van de Fliert and Braun 2002; Gafsi et al. 2006).

Assessing the sustainability of farming systems represents a process of making operational the concept of sustainability. It is a key issue for the implementation of policies and practices aiming at revealing sustainable forms of farming systems. A great number of studies have attempted to develop methodological frameworks for the assessment of sustainability of farming system (Ness et al. 2007; Sadok et al. 2007; Bockstaller et al. 2009). Many studies propose to measure sustainability by the means of a set of indicators. These methods start from the three dimensions of sustainability (economic, social and environmental). Each dimension is broken down into several components, which are identified and selected. Within each component, one or more indicators are defined and measured. For example, economic sustainability can be measured by global agricultural revenue per unit of family labour but also by some other indicators representing capital efficiency, financial autonomy and specialisation (Vilain 2008).

Indicators are usually used in aggregate form at the levels of components or dimensions. The assessment of sustainability of farming systems then involves identifying meaningful components and indicators for each dimension of sustainability and finding a single system scoring that would allow combining these indicators and components into aggregate sustainability measures. The aggregation process, related to integrated and holistic approach of sustainability, leads us to look for a compromising solution, which may bring a balance among different dimensions of sustainability. Most holistic approaches of sustainability assessment give the same weight to the three dimensions. But differences between these approaches appear in the choice of components and indicators and the weight given to different components.

12.2.2 Organic Farming Principles

Organic farming is defined as a form of agriculture, which does not use chemical inputs in its production process, and enhancing the biological and ecological processes to promote soil fertility and good health of animals and plants. It involves holistic view and relies on ecological processes, biodiversity and cycles adapted to local conditions rather than the use of external inputs with adverse effects. It aims also to promote fair relationships and a good quality of life for all involved. According to IFOAM,[2] the basic principles of organic farming are:

The principle of health: organic farming should sustain and enhance the health of the soil, plant, animal and human as one and indivisible.

[2] http://www.ifoam.org/about_ifoam/principles/index.htm (accessed in January 2010).

The principle of ecology: organic farming should be based on living ecological systems and cycles, work with them, emulate them and sustain them.

The principle of fairness: organic farming should build on relationships to ensure fairness at all levels and to all parties – farmers, workers, processors, distributors, traders and consumers.

The principle of care: organic farming should be managed in a precautionary and responsible manner to protect the health and well-being of current and future generations and the environment.

These principles are declined then in more specific goals and targets like the focus on organic matter in soil, the diversity and length of rotations, complementarities between legume crops and cereals as well as between crops and livestock, adaptation of species and races to local conditions, linkages between farmers and consumers, contribution to the local dynamics, etc.

12.3 Design of Method for Sustainability Assessment of Organic Farming Systems

12.3.1 Indicator-Based Method for the Sustainability Assessment of Organic Farming

We opted for an integrated and holistic approach which includes the three dimensions of sustainability: economic, socio-territorial and agro-ecological dimensions. Since the range of sustainability components associated with farming system, within these three dimensions, is potentially very wide, the selection of specific components that would be included in our method was a delicate issue. In order to select components of sustainability common to the organic farming systems and conventional systems, we are based on available indicator-based methods for the assessment of sustainability. Eight indicator-based methods were examined for this purpose. Then for the other components specific to the organic farming systems, we have built it based on the principles of organic farming and respecting the overall coherence of agricultural sustainability. We selected for each dimension four global components (Fig. 12.1).

Economic sustainability dimension includes components relevant to economic situation of the farming system and its ability to continue in the long term. This dimension is not a specific one of organic farming systems, so we used the same components given by the IDEA method (Zahm et al. 2004; Vilain 2008). First of all, the farm should be profitable without taking economic risk to be sustainable. So the viability component includes two aspects: the agricultural revenue per unit of labour and the farm's specialisation. The farm should also have a degree of autonomy from the debt and also in relation to subsidies and public aids, particularly the Common Agricultural Policy (CAP) payments. One key of farming system sustainability relies in the economical transmissibility of the farm. The farm will be easily transmitted

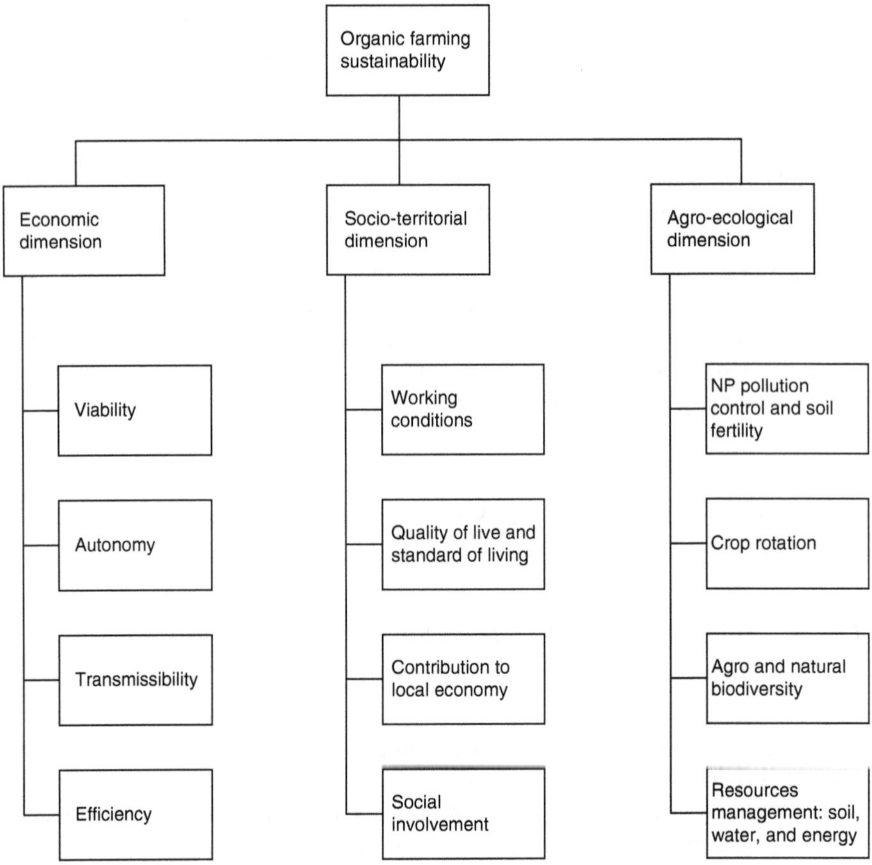

Fig. 12.1 Components of agricultural sustainability

whereas it has a small economic size (asset value per unit of labour), while large and flourishing farms disappear at the end of farmers' career. Finally, the farm will be more sustainable if it increases its farming system efficiency, which will be measured by gross results per capital. Good efficiency means that the farm gets more autonomy from providers which would make it less vulnerable to external market fluctuations.

Both social and ecological dimensions are more specific to organic farming systems. So we have selected components according to sustainable agriculture principles and organic farming principles.

Social dimension of sustainability views farmers in two distinct roles: as producers and members of local society. It is then common to distinguish two types of social dimensions (Gafsi et al. 2006; Van Calker et al. 2006; Vilain 2008): internal social sustainability and external social sustainability. The internal one deals with work conditions within the farm, the quality of life enjoyed by the farmer and the living standard offered by the farming system for farmer and his family.

The external social dimension relates to the farming system's contribution to the sustainability of its territory and community. This contribution would be in the economic and the social levels and consists of a number of different aspects. Contributions to local economy include permanent and seasonal jobs created on farm, prospects of transmission and continuity of farm, direct marketing contributing to the local economy and creating additional jobs, and providing more services in the context of multifunctionality (service provided to local communities, agro-tourism, teaching farm, etc.). For example, on-farm agro-tourism activities may have some positive impact on local communities and also provide an additional income for farmer. Social involvements comprise farmer's participation and responsibilities taken in local organisations, regular contact with consumers because people not only consume agricultural products but are also involved in their production, participation to local and professional networks which lead to sharing experiences and participation to collective actions on joint working, investment group or collective marketing of agricultural products.

Agro-ecological dimension of sustainability examines the propensity of the farming system to combine efficient use of natural resources and minimal environmental cost. It measures the ability of farms to be more or less autonomous in relation to the use of energy and non-renewable resources. But what is also important, according to the principles of the organic farming, is the farm's ability to use and improve agro-ecological complementarities of different productions respecting the balance of the ecosystem. So if it is accepted that the organic farming system causes little risk of pollution, it remains that the great challenge of this system is to better utilise internal resources and agro-ecological balances. The first two components address this issue, particularly the management of soil fertility using organic matter, diversity and length of crop rotations and the use of leguminous plants in rotations. Agro-ecological sustainability requires also a particular attention to both agro and natural biodiversities at planting hedges which are safe havens for zoophagous insects and predators, plant and animal diversity, and adaptation of species and races to local agro-ecological conditions. Finally, the sustainability of farm depends on farmers' practices concerning resources management, particularly the soil, water and energy management.

The next step was to select an appropriate list of indicators for each component. We are based on the existing indicator-based methods to select relevant indicators. Each indicator receives a mark. The sum of marks for various indicators in one component constitutes the global mark for it. By this way, we can estimate the relative weights of different components in the overall sustainability measure. All components do not have the same weight (total of mark), but the three dimensions have the same weight. The sum of marks for different components in one dimension must be 100 points. In order to get this balance between sustainability dimensions, we had set a threshold and scoring system for each indicator.

The final mark for the overall sustainability is the limiting factor, i.e. the lowest among the three dimension marks. It is important to underline that the dynamic orientation towards sustainable agriculture is to be undertaken through three dimensions – economic, socio-territorial and agro-ecological – simultaneously. Being successful on only one dimension is not enough to reach sustainability. We cannot compensate for the weakness of one dimension by good marks in others.

12.3.2 Survey Design and Case Study

To illustrate possible applications of the proposed method, we studied the sustainability of selected farms in the region of Midi-Pyrenees. To do this, we developed a questionnaire to gather information necessary to assess the sustainability of farms, i.e. to measure each indicator and give it a mark. The questionnaire consists of two sections. At the beginning, general information about the farmer (age, education, professional experience, etc.) and farming system (size, types of productions, marketing modes, etc.) are asked. The next section was designed to extract information necessary to various sustainability indicators. The questionnaire is designed to being administrated during an interview with the farmer.

Farms are chosen in the Midi-Pyrenees region, in south-western France, a large region with a diversified agriculture: cow and sheep breeding, crops, mixed farming, fruit and vegetable production, wine making, etc. Except for small very fertile natural regions, the agriculture of Midi-Pyrenees is not very productive. The environment (soils, climate, slopes, etc.) makes the intensification more difficult here than for other French regions, either for crops or breeding. These characteristics and constraints are also valid for organic farming. Data has been collected from 15 farms. The sample has not been made in the aim of being a strict representativity of organic farming in Midi-Pyrenees, but we focused on main production systems (crops, breeding and mixing farming) and introduced variety through the size of farms, duration in organic farming and marketing modes. The farming area is from 34 to 210 ha. Four farms mainly breed cattle, 8 have crops, and 3 are in mixed farming. Half of farms have direct marketing modes. Only three farmers had converted to organic farming before 1999 (this date symbolises the undertaking of French procedure of sustainable agriculture, which encourages the conversion to organic farming).

12.4 Application of the Method

Here we aim to demonstrate how our proposed method could be used and to assess its performances. First, the application of this method has proved easy to implement. The 2-h interview with the farmer enough to get information required all indicators and to measure then the sustainability of his/her farm.

Then, the results of the 15 farms studied show that the proposed method has good sensitivity which allows to observe the differences in sustainability between farms (Fig. 12.2). The overall sustainability mark varies from 27 to 56 and corresponds to the lowest mark among the three dimensions' marks: economic, socio-territorial and agro-ecological. This criterion is important in assessing the performance of a

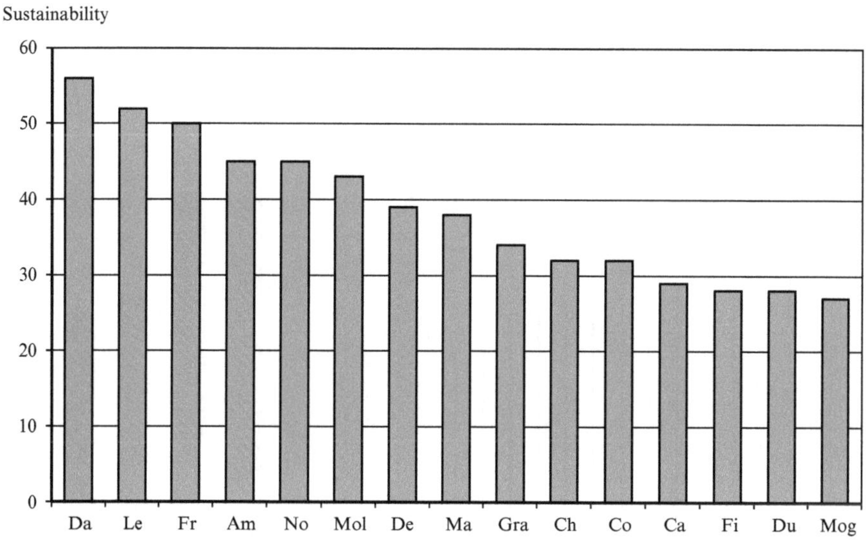

Fig. 12.2 Sustainability rates in the farms surveyed

measurement tool, since it captures the small differences between farms in the process of improving sustainability. Moreover, these differences are not only captured between different production systems but also within the same production system. Figure 12.3 shows the change in rating sustainability in three different systems: crops, breeding and mixing farming. As can be seen, there is considerable variability within each of the three systems.

The proposed method can reflect quite accurately the differences in farming practices. A radar presentation can view synthetically these differences on the 12 components of sustainability. The two farms presented belong to the same system of production as the mixed system, but they have different levels of sustainability (Fig. 12.4). Farm Da has a high area of sustainability compared to the average of all farms. By contrast, farm Mog has a low sustainability area. For this farm, except the components of viability and NP pollution control-soil fertility, all other components are at levels below the group.

The results show also that the methodological work takes into account very well the agronomic specificity of organic farming in terms of integrated approach, soil fertility, long rotations and diversification. Farmers who use many agro-ecological complementarities in their farming systems, by opting for long rotations and the improvement of technical aspects without systematic use of organic fertilisers, have obtained good marks in the agro-ecological sustainability. However, who have opted to simplify their farming systems by choosing a pattern of conventional agriculture (specialisation, short rotations, heavy use of inputs) had a low level of sustainability. Similarly, regarding the socio-territorial sustainability, farmers who have direct contact with consumers and are well integrated into local networks have important marks.

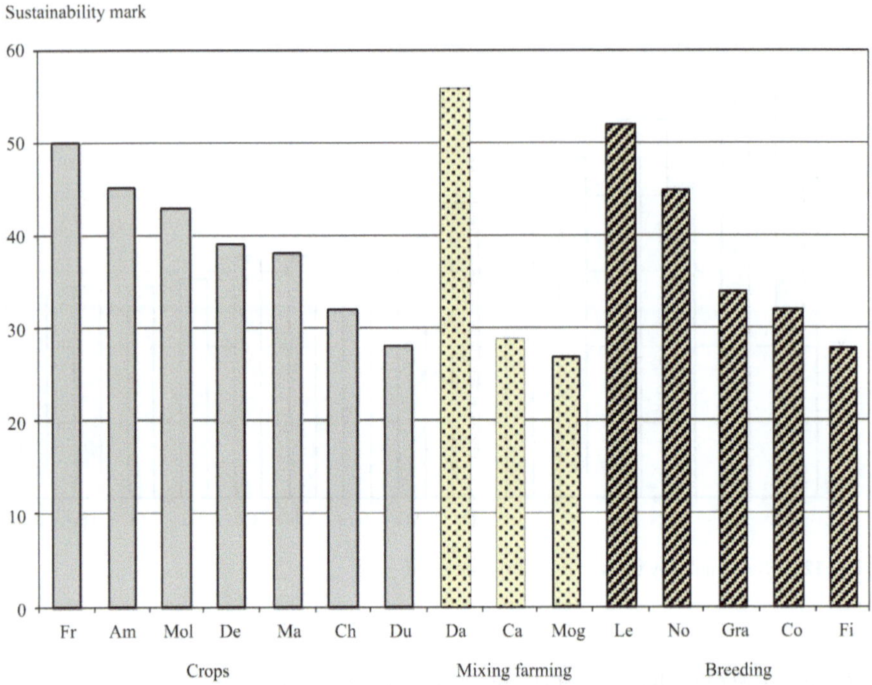

Fig. 12.3 Sustainability rates ordered per production systems

An overall look to sustainability rates in all farms shows a fairly low to medium level. Scores range from 27 to 56. But this reflects the lower value of the three components. Farms have obviously higher values for other dimensions. For example, farm Co has a sustainability score of 32, which corresponds to the note of the economic dimension. This farm has good marks in socio-territorial and agro-ecological dimensions, respectively, 71 and 63. The lowest rating corresponds in 9 cases out of 15 to note the economic sustainability, in 5 cases to note the agro-ecological sustainability and for one case to the socio-territorial sustainability. The economic sustainability is the one in which the grading varies the most (28–65) and is also the worst (average, 42). Only four farms are above average. Generally, on average, the farms in the sample have a weak economical viability (income per worker) and financial autonomy due to debt; they are rather dependent on public subsidies (notably CAP) and quite specialised but are easy to pass on and have good production efficiency.

12.5 Discussion and Conclusions

Taking into account the specificities of the organic farming systems is very important for the development of an effective sustainability assessment tool for such systems. In this chapter, we propose an indicator-based method dedicated to

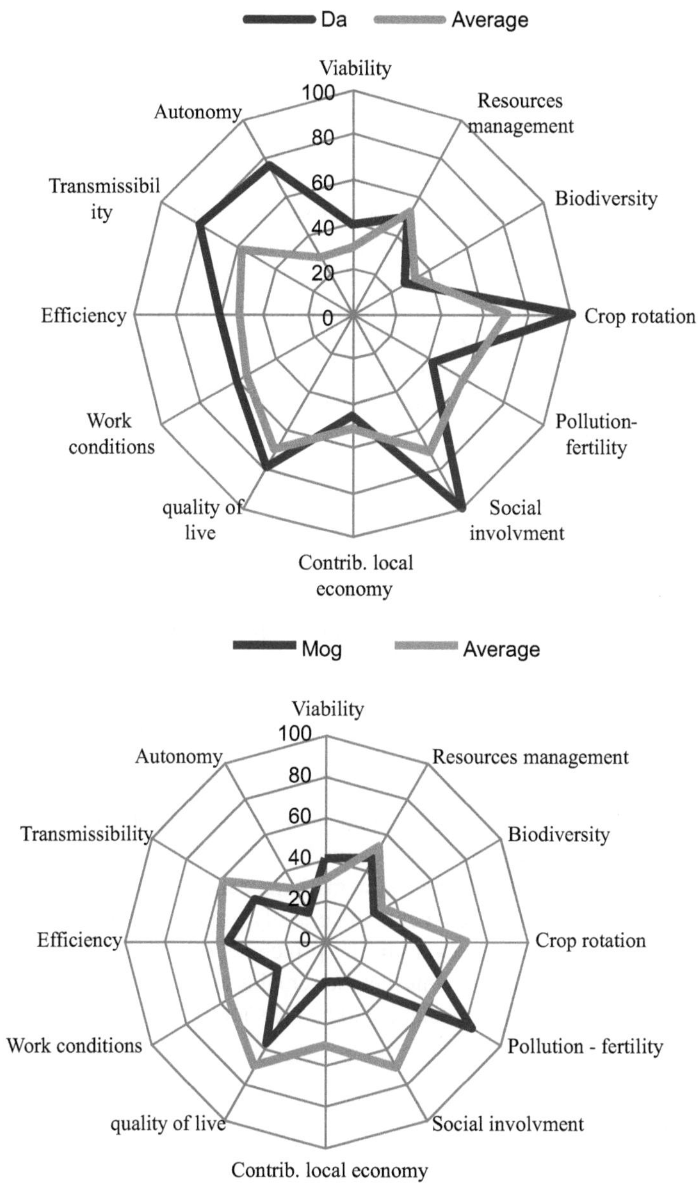

Fig. 12.4 Radar presentation of sustainability rates in two farms

treatment of these specificities. The results of its application in selected farms in Midi-Pyrenees region show the relevance of this method, particularly in the social and ecological dimensions where there is a need to have indicators suitable for organic farming.

Thus, for the ecological dimension, the issue of autonomy of the production system and its consistency with the characteristics of the ecological system has been well identified in the farms and well measured by the indicators. This is a crucial issue for the organic farming that goes beyond the environmental concerns (environmental protection and natural resources) to associate strongly with the agronomic aspects and the productive dimension of agricultural system. For this reason, we chosen to call this dimension "agro-ecological" and not only "ecological" what is usual in the sustainability literature. It is also in the same spirit that we have chosen the name of "socio-territorial" dimension instead of "social" dimension. Indeed, the basic principles of organic agriculture encourage farmers to have a high involvement in the territorial and local level. This occurs through active participation in local networks, the principle of fairness and contact producers-consumers. The assessment method proposed offers indicators dedicated to this topic. The results illustrate the practices of farmers in this area, resulting in a higher mark of the socio-territorial sustainability.

Overall, we can have a degree of satisfaction with the relevance of this method, but improvements are required in terms of taking into account the variability of production systems in organic farming. These systems, in Midi-Pyrenees region as in France, present a very large variability ranging from market-garden crops and arboriculture to cropping systems or livestock systems. Indeed, all the farms surveyed have common production systems (crop, breeding, mixing farming). But specific production systems such as market-garden crops or fruit growing and viticulture have not been studied. But these systems require many adjustments, particularly in agro-ecological indicators.

In conclusion, we can say that the current development of organic farming requires backup from scientists, policymakers and other stakeholders to facilitate the evolution of these farming systems towards sustainable paths. The built of a method for assessing the sustainability of these systems is an action undertaken in this regard. This method has the required elements to be easily used by actors. It is a diagnostic method that is easy to use, relevant to the principles of sustainability and organic farming and synthetic. However, this method remains to be validated in other contexts and other production systems, which could be the subject of further research.

References

Agence Bio. (2009). *L'agriculture biologique en France: Chiffres clés 2008*. Retrieved December 2011, from www.agencebio.org

Andreoli, M., & Tellarini, V. (2000). Farm sustainability evaluation: Methodology and practice. *Agriculture, Ecosystems & Environment, 77*, 43–52.

Bockstaller, C., Guichard, L., Keichinger, O., Girardin, P., Galan, M. B., & Gaillard, G. (2009). Comparison of methods to assess sustainability of agricultural systems: A review. *Agronomy for Sustainable Development, 29*, 223–235.

Gafsi, M., Legagneux, B., Nguyen, G., & Robin, P. (2006). Towards sustainable farming systems: Effectiveness and deficiency of the French procedure of sustainable agriculture. *Agricultural Systems, 90*, 226–242.

Godard, O., & Hubert, B. (2002). *Le développement durable et la recherche scientifique à l'INRA. Rapport intermédiaire de mission*. Paris: INRA Edition.

Hansen, J. W., & Jones, J. W. (1996). A systems framework for characterizing farm sustainability. *Agricultural Systems, 51*, 185–201.

Heller, M. C., & Keoleian, G. A. (2003). Assessing the sustainability of the US food system: A life cycle perspective. *Agricultural Systems, 76*, 1007–1041.

Ikerd, J. E. (1997). *Understanding and managing the multi-dimensions of sustainable agriculture*. Presentation held at the Southern Region sustainable agriculture professional development program workshop, SARE regional training consortium, Gainesville, FL, January 15, 1997.

Kropff, M. J., Bouma, J., & Jones, J. W. (2001). Systems approaches for the design of sustainable agro-systems. *Agricultural Systems, 70*, 369–393.

Ness, B., Urbel-Piirsalu, E., Anderberg, S., & Olsson, L. (2007). Categorising tools for sustainability assessment. *Ecological Economics, 60*, 498–508.

Pacini, C., Wossink, A., Giesen, G., Vazzana, C., & Huirne, R. (2003). Evaluation of sustainability of organic, integrated and conventional farming systems: A farm and field-scale analysis. *Agriculture, Ecosystems & Environment, 95*, 273–288.

Park, J., & Seaton, R. A. F. (1996). Integrative research and sustainable agriculture. *Agricultural Systems, 50*, 81–100.

Rigby, D., Woodhouse, P., Young, T., & Burton, M. (2001). Constructing a farm level indicator of sustainable agriculture practice. *Ecological Economics, 39*, 463–478.

Sadok, W., Angevin, F., Bergez, J. E., Bockstaller, C., Colomb, B., Guichard, L., Reau, R., & Dore, T. (2007). Ex ante assessment of the sustainability of alternative cropping systems: Implications for using multi-criteria decision-aid methods. *Agronomy for Sustainable Development, 27*, 1–12.

Sydorovych, O., & Wossink, A. (2008). The meaning of agricultural sustainability: Evidence from a conjoint choice survey. *Agricultural Systems, 98*, 10–20.

Van Calker, K. J., Berentsen, P. B. M., Romero, C., Giesen, G. W. J., & Huirne, R. (2006). Development and application of multi-attribute sustainability function for Dutch dairy farming systems. *Ecological Economics, 57*, 640–658.

Van de Fliert, E., & Braun, A. R. (2002). Conceptualizing integrative, farmer participatory research for sustainable agriculture: From opportunities to impact. *Agriculture and Human Values, 19*, 25–38.

Van der Werf, H., & Petit, J. (2002). Evaluation of the environmental impact of agriculture at the farm level: A comparison and analysis of 12 indicator-based methods. *Agriculture, Ecosystems & Environment, 93*, 131–145.

Vereijken, P. (1997). A methodical way of prototyping integrated and ecological arable farming systems (I/EAFS) in interaction with pilot farms. *European Journal of Agronomy, 7*, 235–250.

Vilain, L. (2008). *La méthode IDEA: indicateurs de durabilité des exploitations agricoles: guide d'utilisation*. Dijon: Educagri Editions.

Webster, J. P. G. (1997). Assessing the economic consequences of sustainability in agriculture. *Agriculture, Ecosystems & Environment, 64*, 95–102.

Zahm, F., Viaux, P., Vilain, L., Girardin, P., Mouchet C. (2004, November). *La méthode IDEA: une méthode de diagnostic pour passer du concept de durabilité à son évaluation à partir d'indicateurs*. (Paper presented at Partnership for European Environmental Research – PEER Conference, Helsinki).

Zander, P., & Kachele, H. (1999). Modelling multiple objectives of land use for sustainable development. *Agricultural Systems, 59*, 311–325.

Part V
Decision Support Methods for Sustainable Farming Systems

Chapter 13
X-farm: Modelling Sustainable Farming Systems

Alvaro Rocca, Francesco Danuso, Franco Rosa, and Elena Bulfoni

Abstract The aim of this chapter is to illustrate the structure of *X-farm*, a model to manage farming systems under energetic, economical and ecological perspectives, using the dynamic simulation approach. The structure of *X-farm* is composed by some integrated modules representing the main centres of farming costs and production: soil management, crop production and processing and energy production and administration. The dynamic simulation is addressed to find the best combination of crop and livestock activities in the farm plan. The objective of energy production is afforded by using crops and reducing the energy use by optimising energy-saving techniques; the ecological objective is formulated by accounting the CO_2 emissions; the economic objective is targeted to profit maximisation, constrained by the level of achievement of the energy and ecology targets. The dynamic simulation is expected to help in improving the farm management performance with the simultaneous achievement of the three objectives. Finally, combining the *X-farm* model with GIS techniques, the analysis will be expanded to the agro-district planning to support the regional strategy for agro-energy production.

Keywords Farm • Model • Decision-support system • Bioenergy • Scenario • Simulation • Management

A. Rocca (✉) • F. Danuso • E. Bulfoni
Department of Agricultural and Environmental Sciences, University of Udine,
Via delle Scienze 208, 33100 Udine, Italy
e-mail: alvaro.rocca@uniud.it; francesco.danuso@uniud.it; elena.bulfoni@uniud.it

F. Rosa
Department of Food Sciences, University of Udine,
Via delle Scienze 208, 33100 Udine, Italy
e-mail: rosa@uniud.it

A.A. Marta-Costa and E. Silva (eds.), *Methods and Procedures for Building Sustainable Farming Systems: Application in the European Context*,
DOI 10.1007/978-94-007-5003-6_13,

13.1 Introduction

Agricultural researchers widely recognise the importance of sustainable agricultural production systems and the need to develop appropriate methods to measure sustainability (Byerlee and Murgai 2001; Pacini et al. 2003). Bioenergy production efficiency at farm level is still questionable, depending on the commodity used, agronomic practices, climate variability and other unpredictable events. Some researchers assess that the energy balance is still negative (Pimentel 2003; Pimentel and Patzek 2005); other studies (Hill et al. 2006) suggest that the energy produced with the oil and co-products by using energy-saving techniques is significantly higher of the energy spent.

Models are excellent tools to organise knowledge and help to explore alternative scenarios for the management of agricultural systems (Bechini and Stöckle 2007). Farm simulation modelling is assuming increasing importance; oriented to provide short- and long-term scenarios (Danuso et al. 2007), it can be a useful tool to improve the planning capability of the agro-energy farm. Examples of the application of the simulation approach are the whole-farm dynamic model (GAMEDE; Vayssières et al. 2009), integrated farm system model (Rotz and Coiner 2006), FARMSIM (Van Wijk et al. 2006) and SIPEAA (Donatelli et al. 2006).

The increasing complexity from the cropping system to the farming system involves many new fundamental methodological issues for its representation, in particular, the competition among different farm activities for farm resources (manpower, energy, machinery, time window for tillage, etc.). Moreover, the need to simultaneously manage many different fields and different crop rotations creates further difficulties.

In this chapter, *X-farm*, a farm dynamic simulation model developed at the University of Udine (Danuso et al. 2007, 2010), is presented. *X-farm* represents a generic "agro-energy farm", taking into specific account crop biomass production, net energy balance and environmental and economic balances. This farm is targeted to achieve the energetic self-sufficiency, by using a quota of the total biomass produced in farm for the production of energy as oil, biogas or heat.

X-farm is formed by different modules describing the farm activities; they can be grouped in different sections: management, production, soil and accountability (in terms of energy, environment and economy).

Simulations of different cropping scenarios have been performed to test the *X-farm* capabilities to simulate complex farming systems to be used as a decision-support tool.

13.2 Methods

13.2.1 Model Implementation

X-farm has been implemented using SEMoLa (Simple, Easy to use, Modelling Language). SEMoLa (Fig. 13.1) is a software application for the development of simulation models and agro-ecological knowledge integration (Danuso 2003) that

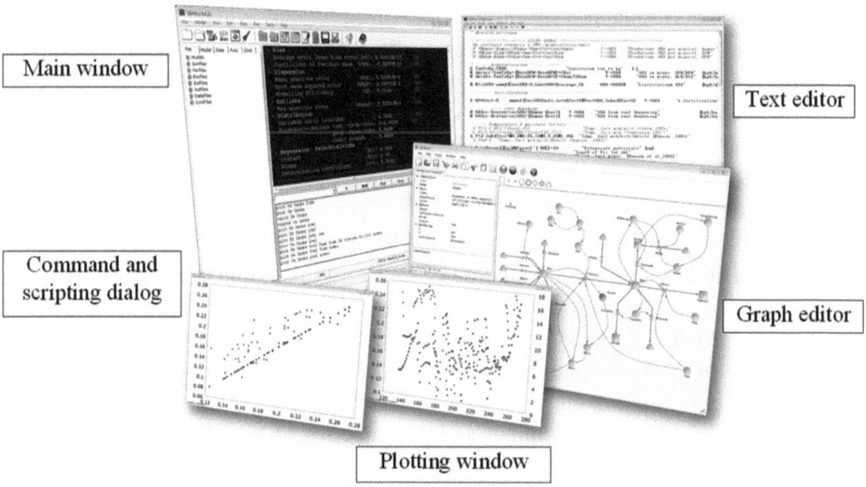

Fig. 13.1 Main dialogs of SEMoLa 6.0 modelling framework

Table 13.1 Graphic representation of the SEMoLa ontology

State	Rate	Impulse	Parameter	Auxiliary variable	Exogenous variable	Event

implements a declarative language. This makes the model code very easy to understand and to modify, even without computer programming skill. Therefore, SEMoLa models can be easily implemented and customised.

SEMoLa has been developed and is maintained at the Department of Agricultural and Environmental Sciences of the University of Udine (Italy). SEMoLa allows the simulation of dynamic systems by the construction of deterministic and stochastic models, based on states (stock and flow) or on elements (individual-based modelling). The ontology of SEMoLa has been inspired by the system dynamics concepts proposed by Forrester (1961), widely used in describing continuous systems (Muetzelfeldt and Massheder 2003).

With SEMoLa language, all farm processes are represented by nine types of concepts (Table 13.1):

1. Material: a quantity that follows the conservation law (conservative quantity). It is opposite to "information" which is not conservative. A farm system can have more than one material (e.g. water, biomass, nitrogen, money, energy), and each material can be in one or more states.
2. Group: an "entity" composed by elements sharing a number of common properties (i.e. state, parameter, etc.). Each element of the group can have its own inputs and outputs. The number of element can vary during simulation by events (e.g. the group of fields, the group of tractors).

3. State: amount of material having specific properties; it evolves in time, thanks to continuous flow (rates) or by sudden modifications caused by events (impulses).
4. Rate: variable that regulates the flow of materials from a state to another or the exchanges of materials from the system and its environment. It depends on system information.
5. Parameter: information of the system, constant during the simulation time. It is a static memory of the system.
6. Auxiliary variable: information obtained from states, parameters and exogenous variables and used in the calculation of rates, impulses and events.
7. Exogenous variable: informative variable generated outside the system and not under the control of the system, able to affect the system itself.
8. Event: something happening that determines sudden modifications of states (by impulses) or parameters.
9. Impulse: variable that determines an instantaneous shift of materials from a state to another, as a consequence of events.

SEMoLa language combines concepts of amount, flow and influence, to usefully describe the interconnected relations in complex systems that increase in complexity when agronomy, ecology, economy and environment are simultaneously considered.

In the *X-farm* model, the farm activities are described with the concepts of state, rate, parameter and event. Crop, livestock and energy productions are also characterised by starting and ending events, temporal windows, priority in accessing resources and prerequisites.

13.2.2 Model Description

At present, the "agro-energy farm" simulated by the *X-farm* model is formed by 23 interconnected modules (Fig. 13.2) logically grouped into four sections: management, production, soil and accounting. The simulation time step is daily.

The farm represented is composed by one or more fields, each one with different soil types, crop rotation and cropping practices. Other simulated activities are cattle husbandry (milk and meat production) in which cows are considered individually, during their productive life. The oil crops can supply seeds for the farm oil extraction chain or for selling to the market.

The *Production* section simulates the crop yield of each field, oil extraction from seeds and milk production from cattle. In particular, crops are represented by the module *CSS-CropYield* derived from the CSS model (cropping system simulator; Danuso et al. 1999); it simulates crop biomass growth and yield under different conditions, depending on climate, soil characteristics, manure and fertiliser applications, tillage and other management choices like irrigation. Potential crop growth is simulated by an implementation of the SUCROS model (van Laar et al. 1997), while phenology and the factors limiting production are implemented as in CropSyst (Stöckle and Nelson 1994). The *XF-Oil* module deals with the oil production

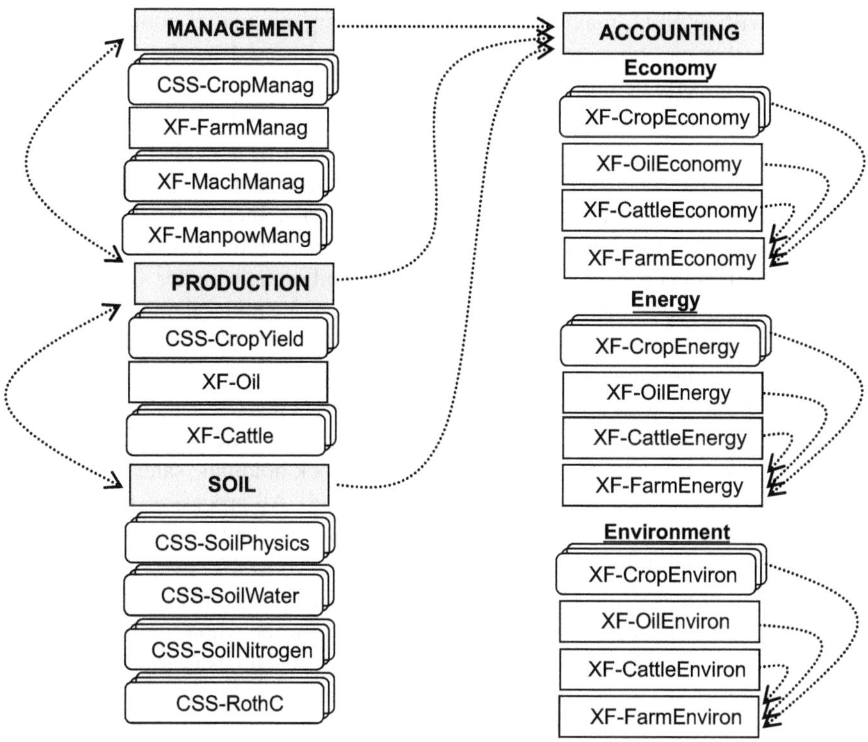

Fig. 13.2 The modules of *X-farm*. *Arrows* indicate the informative relationships among modules. Note that there are two types of modules: simple modules and multiples modules. Multiples modules are represented by the concept of group (individual base model). For example, in the farm, there is only one oil module, but the crop and soil modules are replicated in order to represent each field of the farm

process, consisting of mechanical extraction by seed pressing. This oil can be used as fuel in farm machinery, in cogeneration of electric and thermal energy, or for the production of biodiesel by transesterification. In this way, the energy self-sufficiency of the farm is achieved, and the exceeding oil or energy can be sold to the market (Rosa 2008, 2009). In the *XF-Cattle*, livestock is fed by the cake, being the co-product of the oil extraction and by other feeds from the market. *X-farm* considers the specific conditions of every cow, in terms of age, weight, number of pregnancies and lactation stage. The milk production of each cow is obtained from the specific lactation curve. The co-products, represented by wastes or manure, are spread as organic fertiliser to the fields.

The *Soil* section considers soil as divided into one or two layers, depending on the dynamics of the involved material (water, organic matter, nitrogen). The depth of the upper layer changes according to the crop root growth, from the soil surface to the maximum depth explored by roots during the crop life. The soil type is classified as function of the amount of sand and clay. The other soil characteristics (water field capacity, wilting point, maximum water capacity, organic matter content, etc.) are parameters that can be suggested by the model or inserted by the user. All soil parameters are corrected for the amount of gravel.

The *SoilWater* module simulates the soil water content taking into account actual evapotranspiration, run-off and infiltration. Drainage to water table and capillary rise are simulated, according to Rijtema (1969) and Driessen (1986).

Nitrogen content (as nitrate and ammonium) is calculated in the *Soil Nitrogen* module, separately for root layer and deep layer. Moreover, the model simulates the nitrogen content in crop yields, crop residues and soil organic matter. Crop residues decay is considered in the soil organic matter balance, by an implementation of the *RothC* model (Coleman and Jenkinson 2008). This model divides organic matter into easily decomposable residues, resistant to decomposition residues, humus and microbial biomass, with different mineralisation coefficients.

The *Management* section simulates agricultural cropping activities for each field and farm strategies, related to oil processing, livestock holdings, sales and internal use of products (*XF-cropManag* and *XF-FarmManag*). All processes, requiring the use of resources in terms of manpower and machinery for the farm organisation, are simulated in the modules *XF-ManpowManag* and *XF-MachManag*.

The *Accounting* section is divided in the *economy*, *energy* and *environment*, providing specific balances for crops, oil and cattle and for the whole farming system.

The *Economy* modules calculate the full costs of resources (variable and fixed costs) and revenues for specific farm activities (crops, cattle and oil) and for the whole farm. The profit and economic performance indexes are calculated to provide evidence of the contribution of every activity to the global performance. Economic information, obtained from market prices for agricultural activities (FRIMAT 2008), is used as input parameters to the model. The economic information output is presented as data files to support decisions of investments and the analyses of the performance evaluation of the results obtained in each activity (Rosa 2009).

The *Energy* modules compute both the energy of the farm products and the direct and indirect energy used by crop, oil and cattle production. The Pimentel approach based on transformation coefficients has been used (Pimentel 2003; Venturi and Venturi 2003) in the energy crop module. The parameters for the energy balance in oil processing have been obtained from trials performed at the experimental farm of the University of Udine. Literature data have been used for the cattle energy balance. The information obtained by the energy modules can be used for balance purposes or to estimate the farm EROI (ratio between energy output and input).

The *Environment* modules account for the direct and indirect inputs and outputs between farm and environment. To compare the environmental performance of the different farm activities, an equivalent function for each of them is defined and normalised for LCA (life cycle assessment) approach analysis (Kim and Dale 2005). Information to perform it is obtained from literature and simulated data.

The *X-farm* model is available in two versions:

1. *X-farm* user (XF): the user version, with a reduced number of input parameters and output variables. In this version, most of the model parameters are automatically

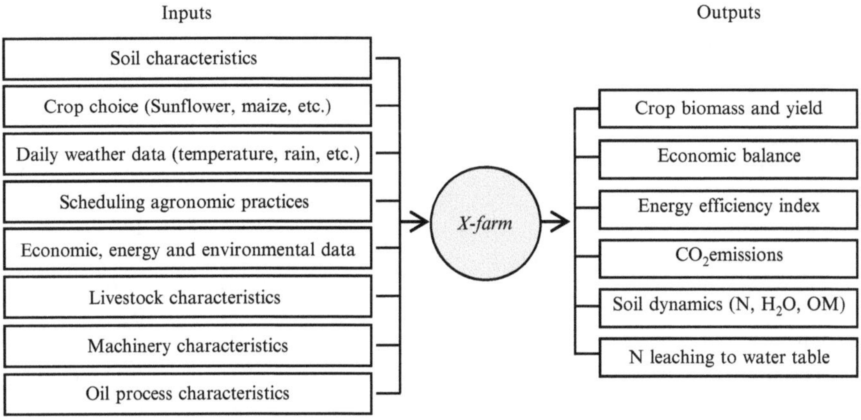

Fig. 13.3 Inputs and outputs for the *X-farm* user model

inserted by selecting a crop, organic fertiliser type, etc. However, the following exogenous input variables are also required: daily minimum and maximum air temperature (°C), rainfall (mm/d), reference evapotranspiration (mm/d) and global radiation at the earth's surface (MJ/m^2/d). This version can be used for farm strategic decision-support and scenario analysis. XF inputs and outputs are reported in Fig. 13.3.

2. *X-farm* development (XFD): the version for modellers, in which all parameters are modifiable and all calculated variables are made available. XFD allows model calibration for specific management situations and can be used as the basis for further model developments.

In the *X-farm* user version, many crop, economic and environmental parameters are built-in to the executable model. In the XFD, they are inserted in files updatable by the user.

13.3 Farm Simulation Experiments

A simulation of the crop production, for different cropping scenarios, performed to show the *X-farm* model capabilities in comparing different farming strategies is presented. As reported in Table 13.2, which summarises the scenarios considered in this application, the *X-farm* model has been run on a hypothetical farm of 100 ha of arable land, using actual meteorological data observed in Udine (north-east Italy, 46°03′N 13°14′E) obtained from the Meteorological Service of the Friuli-Venezia Giulia region, for the period 2000–2003. The cropping scenarios considered involve three crops (maize, soybean and sunflower), 4-year rotations and four fields, differing by land area and soil characteristics. The tillage and other cropping practices are assumed as provided by contractors. Table 13.3 reports detailed information about

Table 13.2 Cropping scenarios for the simulations: soil characteristics and 4-year crop rotation for a hypothetic farm with four fields

	Field 1	Field 2	Field 3	Field 4
Area (ha)	40	25	15	20
Sand (%)	28	40	28	28
Clay (%)	21	19	21	21
Organic matter (%)	3	2.5	3	4
Gravel (%)	5	20	2	18
Soil depth (mm)	1,500	500	1,200	1,000
MWC[a] (mm/mm)	0.40	0.25	0.40	0.40
FC[b] (mm/mm)	0.26	0.10	0.26	0.26
WP[c] (mm/mm)	0.10	0.04	0.10	0.10
2000	Maize	Maize	Maize	Soybean
2001	Soybean	Sunflower	Maize	Maize
2002	Maize	Maize	Maize	Sunflower
2003	Soybean	Sunflower	Maize	Maize

[a] Soil maximum water capacity
[b] Soil field water capacity
[c] Soil wilting point

the events and cropping practices considered in this *X-farm* application example. These practices are based on the techniques usually applied in the north-east of Italy. Irrigation timings and amounts are reported in Table 13.3.

Simulations are set up by preparing a simulation file (*simfile*) that allows to perform simple or multiple simulations. The *simfile* makes a reference to parameters, meteorological data (exogenous variables) and cropping practices (events). Parameters are contained in *parfile*, *gpafiles* and *actfiles*; meteorological data are in *exofile*, and cropping practices are in *evtfile*. They can contain more than one data set that can be selected by customising *simfile*. In this way, it is possible to create different complex simulations combining soil parameters, meteorological data and cropping scenarios.

Parfile contains values for the scalar parameters; *gpafiles* are used to modify values of the group parameters (for fields, cows, etc.), while *actfiles* modify values only when events occur. Parameter values in *parfile* and *gpafiles* are set before the beginning of the simulation. Instead, those in *actfiles* are assigned to parameters at the time of occurrence of specific events (cropping practices).

This structure of input files allows the simulation of different cropping scenarios and crop rotations. Figure 13.4 reports the SEMoLa simulation framework dialogs for editing input files.

Another type of application of the model is the possibility to set up the automatic calculation of irrigation water requirements, in order to maintain the maximum yields but also raising the costs and energy input.

Table 13.3 Cropping practices applied to each crop in rotations

	Harrowing		Mineral fertilisation		Chemical weed control		Planting	Irrigation		Harvest	Ploughing	
Crop	Doy[a]	Depth (m)	Doy	Amount (kg/ha)	Doy	Amount (kg/ha)	Doy	Doy	Amount (mm)	Doy	Doy	Depth (m)
Maize	131	0.15	131	120 P_2O_5	135	2.5	132	176	35	311	102	0.4
								181	25			
								191	35			
			158	90 $N\text{-}NH_4$				200	40			
			184	90 $N\text{-}NH_4$				256	35			
Soybean	131	0.15		-	140	2	150	181	25	300	102	0.4
								191	25			
								200	25			
Sunflower	150	0.15	200	30 P_2O_5	150	2.5	160	181	25	280	102	0.4
								191	25			
			200	80 $N\text{-}NH_4$				200	25			

[a]Day of the year

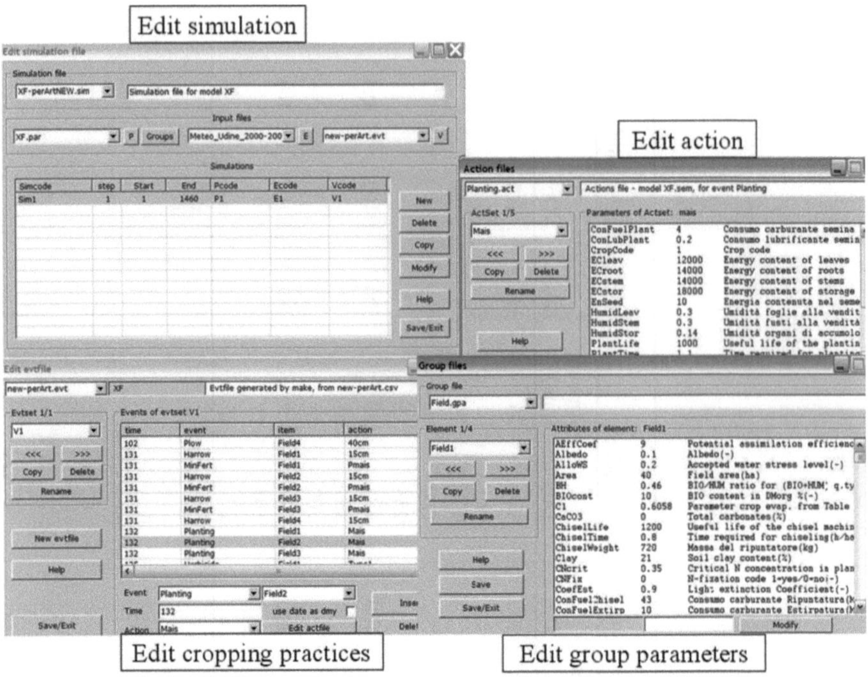

Fig. 13.4 The SEMoLa simulation framework dialogs for editing input files

13.4 Results

Figure 13.5 reports the simulations of biomass accumulation for each crop rotation, over a period of 4 years. These results, obtained by comparing different cropping combinations on a hypothetical farm of 100 ha, provide important information for management decisions in short- and long-term scenarios. The model represents the actual crop production variability that is commonly experienced in the north-east Italy environment. For example, it is possible to observe the stronger effect of the drought on maize yield in 2003 (a year with little rainfall and very high temperatures during the crop cycle). In simulations, we can also detect the effect of the soil type, given that the maize yield differs in fields 1, 2 and 3, in the same year (2000) and with the same cropping practices.

Table 13.4 reports the simulation results of economic and energy accounting. It provides information about the monetary and energy inputs to the farm and about the monetary and energy output obtained from farm activities. These figures can be combined to elaborate a budget and to compare different crops and agronomic techniques, in specific pedological, meteorological and market conditions. The simulation reveals that for almost all the cases, the economic balance of fields and farm results to be only slightly positive. These results, of course, must be interpreted on the basis

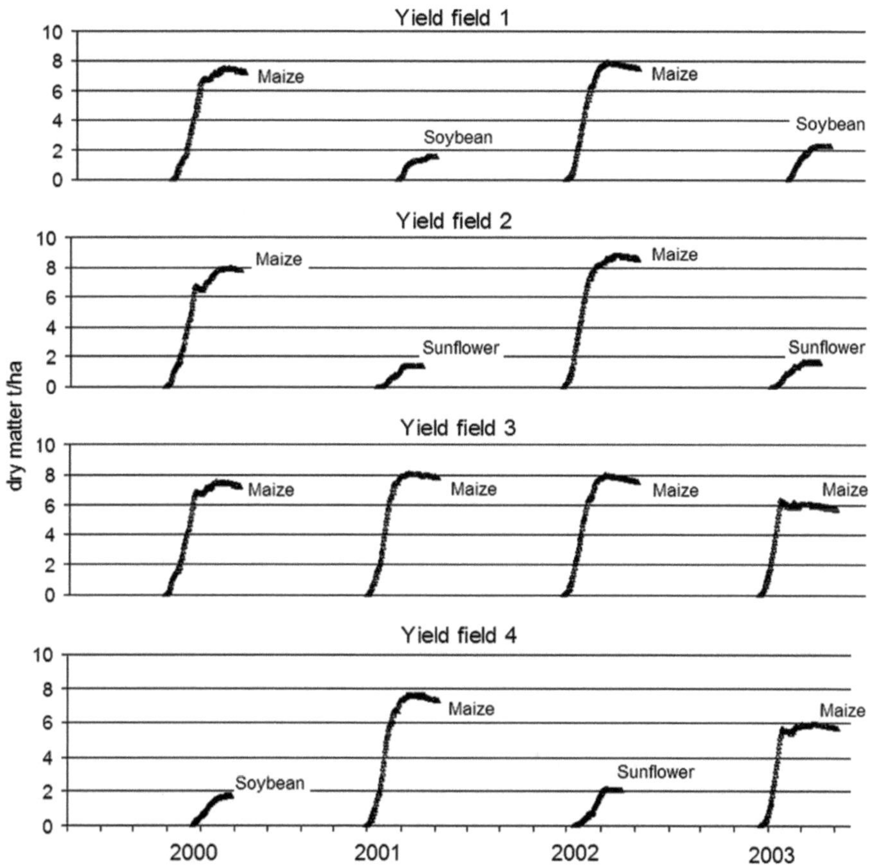

Fig. 13.5 Simulated yields for the four fields of the farm, during the 4 rotation years

of the price levels, cropping scenarios and environmental conditions considered in the simulation trials. The *X-farm* model can therefore be used to explore the effect of different farm management strategies under market and climatic risks. This poor economic result at farm level justifies the introduction of the benefits provided by European agricultural policies, which have not been considered in these simulations. This simulation reflects the actual situations in which farmers' profits are almost equal to the Common Agricultural Policy monetary subsidies.

The energy efficiency, calculated as the ratio between the crop energy output (contained in the total biomass produced) and the direct and indirect energy input (EROI), varies from 5 to 14, with an average value of 6. Among crops, the highest average efficiency has been obtained with soybean. Again, the effect of the bad weather in 2003 generated the worst energy efficiency among years (5.5).

Table 13.4 Economic and energetic accounting of the cropping scenario, for each field and for the whole farm, as simulated by *X-farm*

		Economic accounting			Energy accounting			
		Costs	Revenues	Profit	Input	Output	Balance	EROI
Crop	Year	€/ha	€/ha	€/ha	GJ/ha	GJ/ha	GJ/ha	([b])
Field 1								
Maize	2000	1,074	1,189	115	33	197	164	5.9
Soybean	2001	529[a]	695	166	8	78	70	10.3
Maize	2002	1,110	1,121	11	33	195	162	5.9
Soybean	2003	743	598	145[c]	6	91	85	14.5
	Mean	864	901	37	20	140	120	9.1
Field 2								
Maize	2000	1,155	1,189	34	33	215	181	6.4
Sunflower	2001	377	723	346	14	90	75	6.2
Maize	2002	1,270	1,121	149[c]	33	229	196	6.9
Sunflower	2003	434	723	289	14	92	78	6.4
	Mean	809	939	130	24	156	132	6.5
Field 3								
Maize	2000	1,074	1,189	115	33	197	164	5.9
Maize	2001	1,160	1,121	39[c]	33	207	174	6.2
Maize	2002	1,117	1,121	4	33	197	164	5.9
Maize	2003	853	1,121	268	33	154	120	4.6
	Mean	1,051	1,138	87	33	189	155	5.7
Field 4								
Soybean	2000	581	763	182	8	62	54	7.7
Maize	2001	1,084	1,121	38	33	194	161	5.8
Sunflower	2002	542	723	182	14	118	103	8.2
Maize	2003	842	1,121	279	33	150	117	4.5
	Mean	762	932	170	22	131	109	6.5
		Costs	Revenues	Profit	Input	Output	Balance	
	Year	€	€	€	GJ	GJ	GJ/ha	EROI
Total crop	2000	3,884	4,331	447	110	672	562	6.1
	2001	3,149	3,661	511	90	569	479	6.3
	2002	4,039	4,086	48	116	739	622	6.4
	2003	2,872	3,564	692	89	487	398	5.5
Farm	Mean	3,486	3,910	424	101	617	515	6.1

[a] Soybean field 1, in 2001, received one less irrigation with respect to the other soybean fields
[b] Ratio between energy output and input
[c] Prices of cropping inputs and of crop products are considered the same in the 4 simulation years (at the average level in the last years)

13.5 Conclusions

The *X-farm* model has been presented, and different crop rotations and scenarios on a hypothetic four-field farm have been performed. As highlighted in the simulation outcomes, *X-farm* results to be a useful tool to plan and develop sustainable farming systems. Its use is reasonably simple, and scenario evaluations can be obtained quickly by creating event data files with the agricultural practices and parameters file with the soil traits.

In order to achieve a better description of the farming system, new developments of *X-farm* are currently in progress: (1) biogas production module; (2) implementation of genetic algorithms to obtain robust calibrations and optimisations; (3) improvement of the LCA analysis for different farm energy production; (4) a decision-support system (DSS) version, with the automatic generation of optimised cropping practices decisions (irrigation, automatic generation of mineral fertilisation, ploughing and harrowing events, etc.); and (5) integration between GIS and farm model to create land indicators and to point out trends of specific phenomena (Hartkamp et al. 1999). *X-farm* will be linked to SemGrid (Danuso and Sandra 2006), a raster GIS developed at the Department of Agricultural and Environmental Sciences of Udine University.

Moreover, a major improvement of *X-farm* will be obtained through the implementation (in progress) of the concept of task (activity) in the SEMoLa language. This concept, largely used in operational research, is also going to be adopted in the modelling of farm organisation (Mazzetto and Bonera 2003). The concept of task will allow to deal with (1) management and use of limited resources, (2) agricultural techniques requiring a certain amount of time to be performed and (3) production of by-products, co-products or emissions during the transformation process, operated by the tasks. In SEMoLa, a task is a dynamic process leading to the transformation of the state of a material, which requires the consumption of one or more resources and produces emissions. The beginning and ending of a task is caused by events. For example, ploughing is now treated as an event, instantaneously applied. Considering ploughing as a task, there is a process that transforms the field area from the untilled to the tilled state. This transformation requires resources like fuel, machinery hours, manpower hours, etc. The emissions generated are CO_2 and other pollutants to the atmosphere, heat, etc.

Despite the need for further improvements, the current version of *X-farm* could already be a useful tool to help in planning decisions for agro-energy productions, both at farm and territorial scale.

Both versions are freely available from the authors as an executable file (binary) and also as SEMoLa source code.

Acknowledgements This research has been founded by the project "Valutazione della sostenibilità economica, energetica e ambientale delle filiere agroenergetiche", ex art 17, LR n. 26, 10/11/2005 – Friuli-Venezia Giulia region. The meteorological service of the Friuli-Venezia Giulia region (OSMER) is to be acknowledged for the availability of meteorological data.

References

Bechini, L., & Stöckle, C. O. (2007). Integration of a cropping systems simulation model and a relational database for simple farm-scale analyses. *Agronomy Journal, 99*, 1226–1237.

Byerlee, D., & Murgai, R. (2001 December). Sense and sustainability revisited: The limits of total factor productivity previous term measures next term of sustainable agricultural systems. *Agricultural Economics, 26*(3), 227–236.

Coleman, K., Jenkinson, D. S. (2008). RothC-26.3 A model for the turnover of carbon in soil – Model description and windows users guide. Rothamsted Research Harpenden, Herts AL5 2JQ. ISBA 0 951 445685.

Danuso, F. (2003). SEMoLa: uno strumento per la modellazione degli agroecosistemi. Atti XXXV Convegno SIA. *Napoli, 16–19*(9/2003), 283–284.

Danuso, F., Sandra, M. (2006). SemGrid: Land application of epidemiological and crop models. Pros. IX ESA Congress, 4–7 September 2006, Warszawa, Poland, pp. 631–632.

Danuso, F., Franz, D., Bigot, L., Budoi, G. (1999). CSS: a modular software for cropping system simulation. *Proceedings of the International Symposium "Modelling cropping systems"*, ESA, Lleida, 21–23 June, 1999, Catalonia, Spain, pp. 287–288.

Danuso, F., Rosa, F., Serafino, L., Vidoni, F. (2007). Modelling the agro-energy farm. *Proceedings of the International Symposium "Farming system design 2007"*, 10–12 September 2007 – Catania, Italy, pp 29–30.

Danuso, F., Rocca, A., Andreoni, V., & Bulfoni, E. (2010). Simulation of the agro-energy farm with the X-farm model: Calibration of the crop module for sorghum yield. *Italian Journal of Agronomy, 5*, 3.

Donatelli, M., Acutis, M., Balderacchi, M., Barbieri, S., Bechini, L., Bellocchi, G., Bonera, R., Carlini, L., Danuso, F., Degli Esposti, D., Ditto, D., Fila, G., Fontana, F., Gentile, A., Mazzetto, F., Nasuelli, P. A., Sacco, P., Speroni, M., Trevisan, M., Vetrano, V., & Zuliani, M. (2006). *Modelli per Sistemi Produttivi in Agricoltura*. Bologna: Progetto SIPEAA, CRA-ISCI.

Driessen, P. M. (1986). The water balance of the soil. In H. Van Keulen & J. Wolfs (Eds.), *Modelling of agricultural production: Weather, soil and crops* (p. 76). Wageningen: PUDOC.

Forrester, J. (1961). *Industrial dynamics* (p. 464). Waltham: Pegasus Communications.

FRIMAT. (2008). *Tariffario lavorazioni meccanico-agricole della Toscana*. Roma: UNIMA.

Hartkamp, A. D., White, J. W., & Hoogenboom, G. (1999). Interfacing geographic information systems with agronomic modeling: A review. *Agronomy Journal, 91*, 761–772.

Hill, J., Nelson, E., Tilman, D., Polasky, S., & Tiffany, D. (2006). Environmental, economic, and energetic costs and benefits of biodiesel and ethanol biofuels. *Proceedings of the National Academy of Sciences of the United States of America, 103*(30), 11206–11210.

Kim, S., & Dale, B. E. (2005). Life cycle assessment of integrated biorefinery-cropping system: All biomass is local. In J. Outlaw, K. J. Collins, & J. A. Duffield (Eds.), *Agriculture as a producer and consumer of energy* (p. 319). Cambridge, MA: CABI Publishing Co.

Mazzetto, F., & Bonera, R. (2003). MEACROS: A tool for multi-criteria evaluation of alternative cropping systems. *European Journal of Agronomy, 18*, 379–387.

Muetzelfeldt, R., & Massheder, J. (2003). The Simile visual modelling environment. *European Journal of Agronomy, 18*(3–4), 345–358.

Pacini, C., Wossink, A., Giesen, G., Vazzana, C., & Huirne, R. (2003). Evaluation of sustainability of organic, integrated and conventional farming systems: A farm and field-scale analysis. *Agriculture, Ecosystems & Environment, 95*, 273–288.

Pimentel, D. (2003). Ethanol fuels: Energy balance, economics, and environmental impacts are negative. *Natural Resources Research, 12*(2), 127–134.

Pimentel, D., & Patzek, T. W. (2005 March). Ethanol production using corn, switchgrass, and wood; biodiesel production using soybean and sunflower. *Natural Resources Research, 14*(1), 65–76.

Rijtema, P. E. (1969). Soil moisture forecasting, Note 513. I.C.W. Wageningen, p. 18.

Rosa, F. (2008). A LP model to manage the agro-energy system. *2nd International European Forum on Innovation and System Dynamics in Food Networks 2007*. International Center for Food Chain and Network Research, University of Bonn, Germany.

Rosa, F. (2009). The profitability of Biodiesel with different organization adjustments of the biodiesel chain organization. *4th International European Forum on System Dynamics and Innovation in Food Networks*, Igls-Innsbruck.

Rotz, A. C., Coiner, C. U. (2006). The integrated farm system model – Reference manual, Version 2.0. Pasture Systems and Watershed Management Research Unit – Agricultural Research Service – United Stated Department of Agriculture, 136 pp.

Stöckle, C. O., & Nelson, R. (1994). *The CropSyst user's manual*. Pullman: Biological Systems Engineering Department, Washington State University.

Van Laar, H. H., Goudriaan, J., Van Keulen, H. (1997). SUCROS97: Simulation of crop growth for potential and water-limited production simulations – As applied to spring wheat. *Quantitative Approaches in Systems Analysis,* No. 14, September 1997.

Van Wijk, M. T., Tittonell, P., Rufino, M. C., Herrero, M., De Ridder, N., Giller, K. E. (2006). FARMSIM: A dynamic livelihood model for analyzing management strategies in African smallholder farms. In: Tropentag 2006. International Research on Food Security, Natural Resource Management and Rural Development, Prosperity and Poverty in a Globalized World – Challenges for Agricultural Research. Bonn, Germany, 11–13 October, p. 331.

Vayssières, J., Guerrin, F., Paillat, J. P., & Lecompte, P. (2009). GAMEDE: A global activity model for evaluating the sustainability of dairy enterprises Part I – Whole-Farm dynamic model. *Agricultural Systems, 101*, 128–138.

Venturi, P., & Venturi, G. (2003). Analysis of energy comparison for crops in European agricultural systems. *Biomass and Bioenergy, 25*, 235–255.

Chapter 14
Ecological-Economic Modelling for Farming Systems of Montemuro Mountain (Portugal)

Ana Alexandra Marta-Costa, Filipa Manso, Luís Tibério, and Carlos Fonseca

Abstract This chapter constitutes an assay of a sustainable farming model for the Montemuro mountain (Portugal), a protected area of Natura 2000 network. The model integrates a strategy to reverse the trend of abandonment of this territory, promotes the conservation of natural values through the active management of traditional systems and needs to be economically viable, socially attractive and conducive to more environmental gains. The planning of the farming system was carried having for base the multiobjective programming (noninferior set estimation method complemented with compromise programming). Two objectives were considered to the mathematical formulation of the model: an economic objective – maximising the gross value added – and the second target regarded the environmental scope, minimising energy costs. As results, it is verified that the balance between the selected objectives was established by the selection of a particular set of vegetable activities and always by selecting bovine of trunk Friesian breed and sheep. Bovines of local breed (Arouquesa) are also an option to consider in most of the situations but in smaller numbers.

Keywords Agro-ecosystem management • Ecological-economic modelling • Farming systems • Sustainability • Decision-making

A.A. Marta-Costa (✉) • L. Tibério • C. Fonseca
Centre for Transdisciplinary Development Studies (CETRAD),
University of Trás-os-Montes and Alto Douro (UTAD),
Quinta de Prados, Apartado 1013, 5001-801 Vila Real, Portugal
e-mail: amarta@utad.pt; mtiberio@utad.pt; cfonseca@utad.pt

F. Manso
Centre for Mountain Research, University of Trás-os-Montes and Alto Douro (UTAD),
Quinta de Prados, Apartado 1013, 5001-801 Vila Real, Portugal
e-mail: ftorres@utad.pt

A.A. Marta-Costa and E. Silva (eds.), *Methods and Procedures for Building Sustainable Farming Systems: Application in the European Context*,
DOI 10.1007/978-94-007-5003-6_14,

14.1 Introduction

The Montemuro mountain (PT CON0025) is one of 60 sites included in the National List of Sites of Natura 2000 network adopted by the Council of Ministers Resolution no. 142/97. This place extends over an approximate area of 38,762 ha, including part of the Portuguese municipalities of Arouca, Castro Daire, Cinfães, Lamego and Resende. It is a typical mountainous territory of rugged terrain, crossed by several water courses. Its highest point reaches 1,300 m, and most urban centres are located in an altitude of over 800 m.

The Montemuro mountain is one of the areas classified as more expression in the north of Portugal, not only for the territorial dimension but especially for its role in preserving the landscape and the natural, environmental and cultural heritage of the region. However, this is an uninhabited and aged area, like most regions in the Portuguese inland. Farming in Montemuro mountain site has the following characteristics (*Instituto Nacional de Estatística*, National Statistics Institute – INE 2001, and *Associação de Municípios do Vale Douro Sul*, South Douro Valley Municipalities Association – AMVDS 2008):

- Agriculture is decreasing in Montemuro mountain; it shows a negative evolution regarding both the number of farms with a usable agricultural area (UAA) and the UAA itself.
- The farm size structure is dominated by small-sized farms (3.6 ha in 1999), with a high level of parcelling.
- Undergrowth and uncultivated areas occupy most of the 38,760 ha, whereas the area dedicated to crops is approximately 10% of the total area.
- In view of the zone's specific edaphoclimatic conditions, permanent pastures occupy most of the UAA and are a source of fodder for herds belonging either to local pastoralists or others from further down south or from Estrela mountain.
- Cereal grains, particularly corn and rye and, to a lesser degree, temporary meadows and forage crops, dried leguminous vegetables and potatoes are the most important annual crops, despite having suffered a strong decrease in recent years.
- Vineyards and fruit trees are the most significant permanent crops; apple and cherry orchards (still growing) are also quite important, occupying an area of almost 600 ha.
- The chestnut is one of the few crops which has had a rather positive evolution, although locally this activity has still a limited dimension.
- Livestock, the main source of income for the population in the region, is also decreasing fast. The cattle and small ruminants are the predominant livestock species on the studied area, being the meat of the local cattle breed Arouquesa and of kid goat qualified with protected designation of origin (PDO) and protected geographical identification (PGI).

This study constitutes an assay of a sustainable farming model for the Montemuro mountain (Portugal), integrated in project assigned "Integrated Management Plan of PTCON0025 Montemuro Site". This was developed by the University of Trás-os-Montes and Alto Douro (UTAD), which includes an action plan in the thematic areas of rural development, nature conservation and economic competitiveness, in disadvantaged areas of the mountain.

The developed study aims to identify a number of proposals, given different scenarios, to implement a trial project for a management model of a Natura/Mountain system type. The model must integrate a strategy to reverse the trend of abandonment of such spaces, promote the conservation of natural values through active management of traditional systems and be economically viable, socially attractive and conducive to more environmental gains.

14.2 Methodology

The preparation of a farm plan, following the multidimensions of sustainability, was carried out using the multicriteria decision theory paradigm, as showed by Marta-Costa (2008).

The combined use of different operative techniques of multicriteria decision with the development of mathematical programming models, including technical and economic data characteristics of regional activities, reveal to be "tools" of great significance for the development of systems to support decision-making of managers and farmers (Carvalho 2007). This happens because the process of decision-making in agriculture is a complex procedure that must take into account the different objectives, often in conflict, from the various actors involved (farmers, planners, politicians, consumers).

The concept of agricultural sustainability, integrating environmental, economic and social dimensions, has significantly increased the complexity of decision-making process, given the large number of targets involved and the conflict often generated in its optimisation (Carvalho 2006). It occurs because an increase in the level of performance of one of them may be accompanied by a decrease of others.

The economic competitiveness and environmental sustainability were the main objectives of the delineated model. These dimensions, as referred by Müller (1996), may be considered, in the short term, in conflict, recognising in the long term the interdependence more or less complementary between them. However, it is not possible to attain sustainability through maximising their targets simultaneously. It is necessary to find a balance to achieve it.

In order to arrive at the final farm plan multiobjective programming, in particular, noninferior set estimation method (NISE) complemented with compromise programming was used as methodology.

14.2.1 Mathematical Model

The model, constituted by 135 variables and 103 constraints, was resolved with LINDO – Linear, Interactive, and Discrete Optimizer (LINGO 10 software) – based on the operational aspects of the NISE method and of compromise programming, indicated in Cohon (1978), Zeleny (1982), Romero and Rehman (1989), Romero (1993), Poeta (1994) and Marta-Costa (2008).

The mathematical formulation of a model of an agro-sustainable farm was performed for two different scenarios: with (actual scenario) and without financial support (potential scenario) to the current activities integrated on the Common Agricultural Policy. On a real socio-economic context, the subsidies exist and, for that, are interesting their integration into the model. Moreover, it is assumed that the subsidies may be a situation not sustainable in the long term. Therefore, it is also necessary to provide information about the situation in which financial support is non-existent.

Two objectives are considered: an economic objective – the maximisation of gross value added (GVA), that is, the difference between selling products and buying goods and services, expressed in euros. The situation where the financial support to current activities in the form of subsidies is considered, the value was included in this objective, associated with its supported specific activity.

The second objective reflects environmental considerations – the minimisation of energy costs, expressed in megajoules (MJ), corresponding to the purchase of goods and services to the productive activity. The used energy coefficients were obtained from the reference for energy analysis adopted in the context of the "PLANETE" methodology – "Méthode Pour L'ANalyse EnergéTique de l'Exploitation" (Établissement National d'Enseignement Supérieur Agronomique de Dijon – ENESAD and Agence de l'Environnement et de la Maitrise de l'Énergie – ADEME 2002).

This first objective was selected since a farm's survival requires greater monetary incomes obtained via active participation in the market, that is, the sale of products. The farm profitability is an essential condition for its sustainability, consequently, for the economic development of the region and also a strong contribution for the human fixation in the territory. This objective was translated into the maximisation of the GVA, as this result can easily be processed in the form of a linear equation or inequation.

Regarding the second objective, the intention was that it should reflect the environmental considerations. Thus, among other possible objectives (e.g. minimised water consumption; minimised consumption of pollutant factors of production, fertilisers and crop protection products; minimised use of machines and equipment in the ground), the minimum of energy costs seemed the most suitable given the possibility of quantification of the energy cost in terms of each factor of production used. Even the consumption of water is implicit in this objective through the energy associated with the fuel needed for irrigation (for pump or sprinkler).

This objective was outlined on the point of view that the factor of energy efficiency is an important feature to optimise, in the global economy, being a direct indicator of sustainability. Investigations into several aspects of energy illustrate that the use is generally related to greenhouse gas emissions and to the depletion of natural resources. In order to reduce both effects, potential ways to save energy in farming must be identified (Moerschner and Lücke 2002), thus this issue being the main factor that induced the identification of energy saving as the second goal of farm planning.

Using this approach, it was proposed to improve the economic-environmental conditions of observed farms in the Montemuro mountain, through two deliberately chosen areas: (1) competitiveness in the market with products that present greater GVA and (2) minimal energy costs.

Other goals directly connected to this theme are found in the model, not directly as in the two previous ones, but imposed under the form of restrictions.

The mathematical formulation of the agro-sustainable farm model obeyed, still, few more assumptions that are exposed below.

- The model was constructed based on the information of the geographical area under study.
- The various parameters and technical coefficients were defined according to average characteristics of the farms. Others were based on data published in the literature (INRA 1988; GPPAA 2001; Moreira et al. 2001). Some price levels of inputs and products in producers were obtained by direct inquiry and the others by consultation of prices available on the website of the information system of agricultural markets (http://www.gppaa.min-agricultura. pt/sima.html) and on the available statistics (INE 2005).
- The availability of inputs was defined according to the average characteristics of the farms. For example, the self-owned area of the planned farm was coincident with the average size observed in the universe of farms in the Montemuro area, indicated in official statistics (3.6 ha, INE 2001). It was also considered as a familiar labour force available, the deriving from the elements of the parental household (2 units).
- Two different situations for the use/cleaning of the uncultivated common land were considered. First, the practice of local grazing by farm animals was taken into account. So, in this case, the pastures of the uncultivated common land are just used/cleaned by the local breed of cattle Arouquesa and small ruminants and only during the summer, as indicated by Moreira et al. (2001). In the second situation, it was considered that the uncultivated common land is cleaned using mechanisation. The fodder is cut and given green to the animals of any species present in farm throughout the year.
- On Montemuro farms, potatoes, corn grain, corn fodder, rye, temporary meadows and permanent pastures for hay and for forage morass were identified as main vegetable activities.
- The considered livestock activity is related to the raise of cattle, sheep and goat for meat and/or milk from different breeds. For the first, it was considered adult cows of local breed Arouquesa and Friesian trunk but with situations of descent (F1) of pure animals or resulting from crosses with beef breeds (not pure). Animals from trunk Friesian breeds, situations in which the calves are sold for slaughter to an average age of 9 months, as found for the young descent of Arouquesa cows, and situations in which the cattle are sold at birth were also identified. The sheep and goat rearing has as objective such as the meat, milk and/or cheese production.

- The sale of product activities included the principal products derived from the established vegetables (potatoes, corn grain, rye and hay) and livestock activities (meat, milk and cheese) and also the sale of secondary goods (manure). The first ones constitute the main source of revenue. To develop this model, it was considered that all these crops are sold, including those intended for consumption by the family, the re-employment as a seed or used for animal feed (this is included as a variable charge in the calculation of coefficients of the objective function). The hay is traded only in cases where its production exceeds the needs of the animals. The rye straw is fully used in "beds" of the animals.
- It was considered the existence of provided flow in the market of the main products produced on the farm. Only exception was for the sale of manure because, despite the existence of market for this product, its transaction occurs in limited levels.
- It was also taken into account the possibility of renting land (limited to 25% of self-owner area), hand labour and mechanical traction and the purchase of fertilisers and food for cattle, with the objective that these factors were not restrictive to the expansion of the production process.
- According to practices often developed in the study area and taking into account the environmental conditions, there were established restrictions on the succession of crops and crop rotation. The first was that the intercalary crops (green fodder) succeed or precede, in the same year, the cultivation of potatoes and/or corn. For the latter, it was taken the rotations of rye with potatoes and temporary pasture and also corn in rotation with temporary pasture. It should be noticed that this type of cultural practices have not only a great environmental but also an economic importance, mainly for reasons of fertility and health of crops. In crops, mainly cereals, as indicated by Ferreira et al. (2002), the rotation is especially important to increase the biological fixation of nitrogen and soil organic matter, given the difficulty in obtaining and applying organic fertilisers over large areas.
- On the rational fertilisation context, that is, fertilisation by measure, it is indispensable to obtain the best economic returns from agricultural production and the preservation of the environment quality, namely, the protection of surface and groundwater pollution (eutrophication) (MADRP 1997), and a number of constraints related to the use of fertilisers have been found. These restrictions ensure that the consumption of principal nutrients necessary to vegetable activities is equal or less than the quantity conveyed by manure incorporated into the soil and by synthetic chemical fertilisers bought to the exterior. Nitrolusal 20.5%, foskamónio 7-14-14 and superphosphate of calcium 18% were considered as the most widely used fertilisers, decomposed into their elements (nitrogen, phosphorus and potassium).
- It was yet imposed to the model a reasonable use of nitrogen that does not exceed the amount per hectare specified in the EC's Nitrate Directive (EC 1991). The objective is to protect the underground water from extreme contamination by agricultural nitrates and, in particular, from manure. The amount specified per hectare is the amount of manure that will hold 170 kg of nitrogen (Pau Vall and Vidal 1999). The nutrient content and coefficient of utilisation by crops of

nitrogen from manure of livestock units were obtained from the "Código de Boas Práticas Agrícolas para a Protecção da Água contra a Poluição com Nitratos de Origem Agrícola" (Code of Good Agricultural Practice for the Protection of Water against pollution with nitrates from agricultural sources), published by MADRP (1997).
- Cattle density was ensured to be compatible with the capacity of the natural environment, it was considered that the animal stocking density of farms must be less than or equal to three livestock units per hectare of UAA, in mountain areas, according to the "Good Agricultural Practices" (MADRP 2003).
- The pasture production was estimated according to the values given in the document of Moreira et al. (2001): 12 and 5 tons of dry matter per hectare per year, for pastures more and less productive, respectively.
- The relationship between the production of grain and rye straw and the forecast production for uncultivated common land was obtained from Santos (1991). Although this document refers itself to an area outside the study area (rearing area of Barrosão cattle), it was considered as the approximated values for the Montemuro area, because both of them are situated in mountainous areas, with similar soil, climatic and floristic conditions.

14.3 Results

At this point, the results of the two models delineated for Montemuro mountain are presented, given the assumptions identified earlier. The used techniques allowed to find several solutions for each of the identified scenarios, exposed in this work only the compromise solutions. These solutions belong to the set of efficient solutions that are closer the ideal solution (distance between L_1 and L_∞). Points L_1 and L_∞ define the compromise set. However to choose the options within the set of efficient solutions, belongs to the decision-maker, dependent on preferences attributed to each goal and, consequently, the considered weights in the formulation of the problem. In the present situation, identical weights for each of the objectives were considered.

It should be noted, nevertheless, that given the impossibility of the existence of a non-integer number of animals on farms, provided by the solutions of the initial model, some new compromise solutions were sought (changed compromise solutions). To do so, it was imposed to the initial models the condition that each livestock activity should be equal to the nearest whole number to that obtained with the first compromise solutions.

The changed compromise solutions obtained in the model, in with and without financial support to current activities scenarios, are in Table 14.1.

Observations: (1) For products of activities with different ends, only the portion sold is presented, with the rest reused on the farm (cases of manure and plant products); (2) with the exception of animals for replacement, the remaining born are for sale, as well as refused animals.

Table 14.1 Obtained compromise solutions to the developed models, in with and without financial support to current activities scenarios

	Without financial support		With financial support	
Extreme points	L_1	L_∞	L_1	L_∞
Objectives				
GVA (€)	16,288.871	9,587.9554	18,747.7181	13,222.154
Energy costs (MJ)	246,827.465	146,435.1072	243,666.2572	159,572.1016
Principal decision variables				
Irrigated land (Ha)				
Potato	0.2040	0	0	0.3075
Rye	0.2040	0	0	0
Temporary pasture	0.2040	0.612	0.612	0.3045
Intercalary crops	0	0	0	0.1754
Dry land (Ha)				
Potato	0.6178	0.7379	0.8128	1.2644
Permanent and community pasture (Ha)				
Hay	0.7841	0.7841	1.4832	1.0317
Pasture	1.5919	1.5919	1.5919	1.5919
Community land with grazing by farm animals	5.1492	0.9092	2.9613	1.4012
Cattle (LU)				
Arouquesa (pure F1)	1	0	1	2
Friesian (pure F1 – sale 0 months)	1	0	1	0
Friesian trunk (not pure F1 – sale 0 months)	6	4	6	5
Sheep and goats (LU)				
Sheep (meat and cheese)	30	17	28	9
Crops and animal products sale (kg)				
Potato	9,703.739	8,515.302	9,380.255	18,472.68
Hay	5,513.064	7,240.944	11,386.5	7,406.462
Cow milk	44,550.80	25,457.6	44,550.8	31,822
Sheep cheese	1,080	612	1,008	324
Bovine manure	70,000	39,545.47	70,000	48,131.12
Inputs purchase (kg)				
Dry arable land	0.0058	0.1259	0.2008	0.6524
Irrigated pastures	0	0	0.6992	0.2476
Manpower (hours)	47.8777	0	58.2390	0
N	31.3911	56.6191	57.1562	86.6085
P_2O_5	35.4034	81.6171	42.0593	101.0783
Corn grain	473.9619	0	0	1.4457
Commercial concentrate	37,145.447	20,148.123	35,740.12	20,411.242
Corn silage	33,176.872	27,280.914	35,658.743	18,519.747

The analysis of the obtained solutions allows the following observations:

- The relationship of conflict between the considered objectives is confirmed, since the rise of GVA indicates an increase in its energy costs.
- The selected model activities reflect an accentuated use of the areas by crops connected to the cattle activity. For example, irrigated temporary grassland is an activity always present. The dry arable land is, in general, occupied with potatoes.
- The areas of permanent pastures for hay and for pasture are fully used, as imposed to the models. When the financial supports to current activities are considered, it is necessary to rent irrigated pastures. The pastures from the common lands are used/cleaned directly by the animal grazing, rather than mechanical cleaning. They are generally consumed in a proportional relationship to the cattle identified in the solutions.
- Among the various hypotheses provided to the model for the cattle activities, in all the obtained solutions, the selection of Friesian trunk animals with sales to the birth of their not pure young occurs, in numbers ranging from 4 to 5 animals. Arouquesa cattle should also be present on farms, in numbers between 1 and 2 livestock units, with the exception of one solution. Pure cattle of Friesian are only considered in two solutions, which coincide with those where the economic objective assumes greater values (points L_1). In such solutions, it is also denoted, in general, a higher number of Friesian trunk than the other solutions.
- The model considers, in all compromise solutions, the existence of sheep for meat and milk, being this entirely processed into cheese.
- The farm vegetable products sales, including hay, is observed when the availability exceeds the needs of the animals. This situation occurs mainly by the replacement of those foods for others with high content in protein and with lower volumes of dry matter (commercial concentrate and corn silage).
- Also, manure is sold to the imposed limit, except in solutions where environmental objective is improved.
- The hiring of temporary labour was relatively low, being only required in two situations that coincide with the highest animal density. Noted that while minimising temporary labour is not a clear objective defined in formal model, but identified indirectly in the objective of maximising GVA through its economic cost, its reduction is important due to scarcity of available temporary labour in the region. The surplus of labour in all obtained solutions is also observed but with non-uniform distribution for the periods in question.
- Also with the traction, it is verified that the existing availability is more than sufficient for the needs and is not necessary to hire it.
- The quantities of fertilisers to be acquired to the exterior vary inversely with the amount of manure applied to land.
- The commercial concentrates and corn silage are identified in a proportional relationship with the cattle of Friesian trunk breeds. This should be distributed on a regular basis throughout the year.

14.4 Final Considerations

Two models of agro-livestock farming for two distinct scenarios (with and without subsidies at current activity), in the context of sustainability, are presented in this work. The used techniques allowed the finding of various solutions, and to this extent, the system can be considered open. This means that all of the solutions present advantages and inconveniences, when the results are analysed within possible alternative scenarios with diverse socio-economic circumstances and where goals can have more or less importance.

Based on the obtained extreme compromise solutions, considering equal importance for considered goals, it appears that the balance between them is given to the selection of potato activities (irrigated and dry), temporary grassland (irrigated), marshes and hay grazing and the use of uncultivated common land. Livestock activities are always selecting animals of Friesian trunk (between 4 and 6 animals with not pure descendants and 1 animal with pure progeny) for production of meat and milk and sheep for meat and milk (between 9 and 30 adult animals), the latter is being all transformed, in cheese, in order to ensure the economic performance of the farm. The animals of Arouquesa breed (the local breed) are also an option in almost all situations, ranging from 1 to 2 adult cows.

The development of defined models and its solutions have raised a few points, bearing in mind the framework of the objectives that pretends to be achieved. These are:

- The area of farms is one of the main limitations of results. Increasing the limit of the leased area, for example, the solutions differ from those achieved initially by improving economic performance. Moreover on the hypothesis of food, requirements relating to livestock activities were mainly fulfilled by at least 50%[1] of food produced on farms; this resulted in an absence of any livestock activity, with a penalty in its economic performance.
- There is a large surplus on labour force (in same periods) and in traction. In this sense, for labour force that is not used, it is necessary to find alternative ways for its employment. Likewise, it must be considered the sale or use of machinery and equipment with very high hourly costs in exterior works, to turn it profitable.
- The subsidies level for animals of local breeds does not appear sufficient to encourage their raise, nor to overcome the profits from the animals of exotic breeds.

[1] Condition defined in "standard for beef cattle" in the extensity project – Environmental Management Systems and Sustainability in Agriculture Extensive. Its objective is to get the forage-livestock balance obtained on the farm but also to assure the autonomy of the production unit, fundamental condition to its sustainability (Domingos et al. 2005).

References

AMVDS. (2008). *Plano de gestão integrada do sítio PTCON0025 Montemuro, memória descritiva complementar*. Resende: AMVDS.

Carvalho, N. S. (2006, December). *Agricultura sustentável em regadio e o seu planeamento face aos desafios económicos, sociais e ambientais (o caso da futura zona irrigada de intervenção de Alqueva)*. Paper presented at 5th Iberican congress of water management and planning, University of Algarve, Portugal.

Carvalho, N. S. (2007, October). *O planeamento agrícola face aos actuais desafios ambientais, sociais e económicos (uma aplicação da programação por metas lexicográficas ao caso da futura zona de regadio de intervenção de Alqueva)*. Paper presented at V national congress of Portuguese association of agrarian economy, UTAD, Portugal.

Cohon, J. L. (1978). *Multiobjective programming and planning*. Mathematics in science and engineering, vol. 140. San Diego: Academic.

Domingos, T., Rodrigues, O., Avelar, T., Brito, A., Piçarra, A., Sendim, A., Ferreira, F., Dias, N., Crespo, D., Crespo, J., Lopes, A., Belo, C., Alcazar, R., Sarmento, N., Sequeira, E. (2005). *Sustentabilidade garantida – norma para carne bovino*. Projecto extensity – sistemas de gestão ambiental e de sustentabilidade na agricultura extensiva LIFE ENV/P/505.

ENESAD, & ADEME (2002). *Analyse énergétique d'exploitations agricoles et pouvoir de réchauffement global. Méthode et résultats sur 140 fermes française*. Document du groupe Planète.

GPPAA. (2001). *Contas de culturas das actividades vegetais – modelo de base microeconómica*. Lisbon: GPPAA, CD-ROM.

INE. (2001). *Recenseamentos gerais da agricultura. Dados comparativos 1989–1999*. Lisbon: INE, CD-ROM.

INE. (2005). *Estatísticas agrícolas 2004*. Lisbon: INE.

INRA. (1988). *Alimentation des bovins, ovins & caprins*. Paris: INRA.

Marta-Costa, A. A. (2008). A tomada de decisão no planeamento da exploração agrária no contexto da sustentabilidade: o caso da produção de carne bovina maronesa. Doctoral thesis, UTAD, Vol. I e II.

Moerschner, J., & Lücke, W. (2002). Energy investigations of different intensive rape seed rotations – A German case study. In E. Van Ierland & A. Lansink (Eds.), *Economics of sustainable energy in agriculture* (pp. 27–40). Dordrecht: Kluwer Academic.

Moreira, N., Aguiar, C., Pires, J. (2001). *Medidas agro-ambientais: 3.3 lameiros e outros prados e pastagens de elevado valor florístico - pastagens de montanha*. Lisboa: DGDR.

Poeta, A. M. S. D. (1994). A tomada de decisão no planeamento da exploração agrícola num contexto de objectivos múltiplos. Doctoral thesis, UTAD.

Romero, C. (1993). *Teoria de la decisión multicriterio: conceptos, técnicas y aplicaciones*. Madrid: Alianza Editorial.

Romero, C., & Rehman, T. (1989). *Multiple criteria analysis for agricultural decisions*. Developments in agricultural economics 5. Amsterdam: Elsevier.

Santos, J. M. L. (1991). Sistemas de aproveitamento de recursos naturais no alto Barroso. Doctoral thesis, Technical University of Lisbon.

Zeleny, M. (1982). *Multiple criteria decision making*. McGraw-Hill Series in Quantitative Methods for Management. New York: McGraw-Hill.

Chapter 15
Evaluating Socio-economic and Environmental Sustainability of the Sheep Farming Activity in Greece: A Whole-Farm Mathematical Programming Approach

Alexandra Sintori, Konstantinos Tsiboukas, and George Zervas

Abstract Ruminant livestock farming is an important agricultural activity, mainly located in less favoured areas. Furthermore, ruminants have been identified as a significant source of GHG emissions. In this study, a whole-farm optimisation model is used to assess the socio-economic and environmental performance of the dairy sheep farming activity in Greece. The analysis is undertaken in two sheep farms that represent the extensive and the semi-intensive farming systems. Gross margin and labour are regarded as socio-economic indicators and GHG emissions as environmental indicators. The issue of the marginal abatement cost is also addressed. The results indicate that the semi-intensive system yields a higher gross margin/ewe (179 €) than the extensive system (117 €) and requires less labour. The extensive system causes higher emissions/kg of milk than the semi-intensive system (5.45 and 2.99 kg of CO_2 equivalents, respectively). In both production systems, abatement is achieved primarily via reduction of the flock size and switch to cash crops. However, the marginal abatement cost is much higher in the case of the semi-intensive farms, due to their high productivity.

Keywords Dairy sheep farming • Mathematical programming • GHG emissions • Socio-economic performance • Environmental performance • Abatement cost

A. Sintori (✉) • K. Tsiboukas
Department of Agricultural Economics and Rural Development,
Agricultural University of Athens (AUA), Iera Odos 75, 11855 Athens, Greece
e-mail: al_sintori@yahoo.gr; tsiboukas@aua.gr

G. Zervas
Department of Nutritional Physiology and Feeding Faculty of Animal Science and Aquaculture, Agricultural University of Athens (AUA),
Iera Odos 75, 11855, Athens, Greece
e-mail: gzervas@aua.gr

A.A. Marta-Costa and E. Silva (eds.), *Methods and Procedures for Building Sustainable Farming Systems: Application in the European Context*,
DOI 10.1007/978-94-007-5003-6_15,

15.1 Introduction

Ruminant livestock farming, especially sheep farming, is an important agricultural activity in Greece, since it is mainly located in less favoured areas of the country and utilises less fertile and abundant pastureland. The activity yields income for thousands of farms mainly located in marginal areas, where few alternative economic activities can develop. These farms are dairy farms, since they aim primarily at the production of sheep milk that is responsible for over 60% of their gross revenue and secondarily at the production of meat (Kitsopanides 2006). It is estimated that almost 40% of the total milk produced in Greece is sheep milk. Furthermore, the activity contributes highly in regional development and helps maintain the population in the depressed areas, where it is located. Therefore, the preservation of the activity and the income it yields is important not only for farmers but also for policymakers.

The prevailing sheep farming system in the country is the extensive system, in which the feed requirements of the flock are met mainly through grazing. Extensive breeding farms are characterised by low invested capital with low-productivity flocks, consisting mainly of native breeds (HMRDF 2007[1]). More modern and intensive farms that are also present have a higher invested capital and aim to increase their productivity through supplementary feeding, mainly from on-produced cereals and forage. These two main production systems identified in the Greek sheep farming activity have different characteristics and therefore different economic and environmental performance.

The matter of greenhouse gas (GHG) emissions has recently received extra attention in light of the Kyoto protocol and Europe's commitment to reduce emissions. Agriculture has been identified as a significant source of GHGs, and farmers are urged to adopt not only economically viable but also environmentally sound farming practices. GHG emissions are particularly high in the case of ruminant livestock farming because of methane production through enteric fermentation (Pitesky et al. 2009). The issue of GHG emissions in livestock farms has been addressed in a number of studies that focus mainly in dairy cow and cattle farms (Olesen et al. 2006; Weiske et al. 2006; Veysset et al. 2010). On the other hand, studies that focus on the emission of GHGs from sheep farms refer mainly to meat and wool production farms that have different technicoeconomic characteristics than dairy sheep farms (e.g. Benoit and Laignel 2008; Petersen et al. 2009).

This study aims primarily at the evaluation of the socio-economic and environmental performance of the dairy sheep farming activity in Greece, through the use of a whole-farm optimisation model. In this model, environmental performance is measured through the estimation of the net GHG emissions of the sheep farms. The issue of the GHG abatement cost is also addressed, since mitigation leads to loss of

[1] Hellenic Ministry of Rural Development and Food.

income. The analysis is undertaken in two farms representing the extensive and the semi-intensive farming systems that are commonly found in the country. In the next section, the mathematical model used in the analysis is described in more detail. The characteristics of the extensive and the semi-intensive farms are also presented. The third section contains the results of the analysis, and the final section includes some concluding remarks.

15.2 Data and Methods

Mathematical programming models are commonly used in agricultural studies (e.g. Alford et al. 2004; Veysset et al. 2005; Crosson et al. 2006). They yield the optimal amongst all feasible farm plans, taking into account technical and agronomic constraints of the farms. In the case of livestock and crop-livestock farms, the complexity of the farm operation and the substitution possibilities between alternative activities require the use of a model that can capture all the interrelationships of these activities. The multiple sources of GHGs in crop-livestock farms present another reason for a mathematical programming model to be used (De Cara and Jayet 2000). Thus, a number of studies have utilised mathematical programming models to assess GHGs from various sources and identify cost-effective mitigation strategies (e.g. Smith and Upadhyay 2005; Breen and Donnellan 2009; Petersen et al. 2009).

Therefore, a whole-farm, mathematical programming model is considered an appropriate tool for the estimation of the socio-economic and environmental performance of livestock farms. The model used in this analysis incorporates all livestock and crop activities of sheep farms. The characteristics of the farm model are described in more detail in the following paragraphs. The data used in the analysis is also presented in this section.

The first step of our methodology is to use this mathematical model to obtain the optimal farm plan of each of the sheep farms. This optimal farm plan is derived through gross margin maximisation that is assumed to be the objective of the farmers and is used to measure the economic performance of the farms. Labour inputs in this optimal farm plan are considered as an indicator of the social performance of the farm, and net GHG emissions are regarded as an environmental performance indicator. The second step of our methodology is to estimate the optimal farm plan across increasing levels of abatement and assess impact on gross margin and labour. Following a number of studies (e.g. De Cara and Jayet 2000; Smith and Upadhyay 2005), this is achieved by inserting an additional constraint in the model. Specifically, if e_0 is the original level of net emissions, at the optimal farm plan, and α is the level of abatement ($\alpha < 1$), then a new constraint is inserted in the model not allowing the net farm emissions to be more than $(1-\alpha)\ e_0$. The shadow price of net emissions is also estimated because it indicates the GHG marginal abatement cost for each farm (De Cara and Jayet 2000; Smith and Upadhyay 2005).

15.3 Model Specification

The crop-livestock model used in this analysis maximises total gross margin under the technicoeconomic constraints of the sheep farms and yields the optimal farm plan. For this purpose, it utilises detailed farm-level data on all crop and livestock activities of the farms. The decision variables and the constraints of the model are presented in the next paragraphs. The GHG emission sources that have been taken into account in this analysis are also presented in detail in this section as indicators of the environmental performance of the sheep farms.

15.3.1 Crop and Livestock Activities

Crop activities of the sheep farms involve mainly forage and cereal production for livestock feeding. In the model, farmers can produce cereals and forage either for consumption in the farm or for sale, according to what maximises their gross margin. The two farms used in this analysis produce only alfalfa and maize, which are the main crop activities of the sheep farms of the area where the analysis is undertaken.

Two livestock activities are incorporated in the model, according to the time of sale of the lambs. In the first one, lambs are sold after weaning (approximately 42 days after lambing) and the ewes are then milked. The second activity involves the rearing of the lambs for 3 months prior to their sale. In this second alternative, the live weight of the lambs sold is higher, but the price per kilogram is lower. Also, the milk yield is much lower, since lambs are allowed to wean for a longer period of time. The produced fodder is used for the feeding of the livestock. In the model, there is also a set of variables to approximate monthly distribution of the produced feed. Additionally, monthly consumption of purchased maize and alfalfa presents another set of the model variables. Also, the model includes decision variables that reflect the use of pastureland and the monthly consumption of grass. The final set of variables incorporated in the model involves the monthly labour inputs (family and hired labour inputs in crop and livestock activities).

15.3.2 Feed Requirements

The main component of the model ensures that the monthly feed requirements of the flock are balanced. Minimum intake of dry matter, net energy of lactation, digestible nitrogen and fibrous matter is ensured through monthly constraints. The feed requirements of the flock are estimated according to Zervas et al. (2000). For the productive ewes, these feed requirements include requirements for maintenance, pregnancy and lactation. For the rams, the requirements refer to their maintenance, while for the replacement animals, the feed requirements are estimated every month taking into account the live-weight gain. The weight gain is also taken into account in the case of the lambs, for which feed requirements are estimated for the period that they remain in the farm minus the feed requirements that are satisfied from milk. On-produced feed crops, external feed inputs and available pastureland are used for the balance of the feed requirements of the flock. The nutritional value per kilogram of maize, alfalfa and grass is taken from Kalaisakis (1965) and Zervas et al. (2000).

15.3.3 Labour and Land Constraints

A second component of the model ensures that monthly labour requirements of all production activities are balanced mainly with the family labour inputs. Additional hired labour can be used if necessary in both livestock and crop activities. Labour requirements differ between farms according to the specific crop and livestock activities, management practices, type of machinery used and specific land characteristics. Land availability constraints are also incorporated in the model. They refer to the availability of irrigated land, used for alfalfa and maize production; availability of pastureland; and total farmland.

15.3.4 GHG Emissions

An extra component has been added in this model that refers to the GHG emissions. The main GHG emissions, from livestock farms, are methane (CH_4) from enteric fermentation and excreta and nitrous oxide (N_2O) from excreta. In addition, in a crop-livestock farm, nitrous oxide (N_2O) emissions from fertiliser use should also be accounted for (see, e.g. Schils et al. 2007; Petersen et al. 2009; Veysset et al. 2010). Carbon dioxide (CO_2) emissions from energy consumption pose an additional source of GHGs. In our analysis, all the potential sources of GHGs have been taken into account when total emissions are estimated.[2]

Methane production from enteric fermentation is the most important source of GHGs in livestock farms, and it is associated with the feeding practices of each farm. Farmers choose to feed their flock with on-produced feed and purchased feed taking into account the cost and the nutritional value of each feedstuff and the feed requirements of the flock. The ration used in this analysis is not fixed, but

[2] CH_4 and N_2O have been converted to CO_2 equivalents using the following conversion factors: 1 kg of CH_4=25 and 1 kg of N_2O=298 (IPCC 2006).

it is optimised (see also Petersen et al. 2009). Following the work of De Cara and Jayet (2000), methane emissions are predicted for each feedstuff according to the following equation:

$$E - CH_4/EB = -1.73 + 13.91dE \quad (15.1)$$

where E-CH_4/EB is the percentage share of gross energy loss in methane and dE is a digestibility index, for each feedstuff. The digestibility index is taken from Kalaisakis (1965). Furthermore, the following equation proposed by Vermorel et al. (2008) is used for the estimation of methane emissions from grass consumed by grazing sheep:

$$Y'm = -0.150dE + 21.89 \quad (15.2)$$

where Y'_m refers to methane (in kcal) per 100 kcal of metabolisable energy. Methane produced from livestock excreta is considered negligible, since no anaerobic conditions exist during the management of manure or grazing of sheep (IPCC 2006; Petersen et al. 2009). On the other hand, when aerobic conditions exist, N_2O is produced, and, therefore, direct and indirect N_2O emissions from livestock excreta, deposited on pastureland and managed in piles, are included in the analysis. These emissions are estimated according to the Tier 1 methodology, proposed by the IPCC (2006).

In our analysis, we have included direct and indirect N_2O emissions from the use of nitrogen fertilisers. First, the total amount of nitrogen applied in fields has been calculated using the amount and the type of fertiliser (De Cara and Jayet 2000; Petersen et al. 2009). Then, direct, indirect and leaching emissions from the applied N have been estimated according to the tier 1 methodology and the emission factors proposed by the IPCC (2006). Pre-chain emissions have also been estimated and included in the analysis, following the work of Olesen et al. (2006). As mentioned above, farmers choose whether to feed their flock with on- or off-farm-produced crops. Therefore, emissions from the nitrogen fertilisers used for the off-farm production of feedstuffs have also been estimated and incorporated in the model. Specifically, N_2O emissions from purchased alfalfa and maize have been estimated using data gathered from 85 and 73 farmers of the area, respectively.

Carbon dioxide from energy use is another source of GHG emissions in crop-livestock farms. The main sources of energy in these farms are fuel (mainly diesel) and electricity (see also Olesen et al. 2006). To estimate the emissions from energy use, fuel or electricity requirements for every operation and type of machinery are estimated and multiplied by emission factors (Petersen et al. 2009). As in the case of N_2O, CO_2 emissions from energy requirements of purchased feed are also estimated, according to the data gathered from the farmers of the area. Other inputs like fertilisers and pesticides used in both produced and purchased crops have also caused GHG emissions when they were manufactured. These emissions have been taken into account as well, using farm-level data to estimate the amount of inputs used and related literature to estimate the emissions caused by the manufacture of these inputs. Carbon dioxide emissions from the manufacture of fertilisers are taken from Wood and Cowie (2004), and emissions from the manufacture of pesticides are taken from Audsley et al. (2009).

Sheep farming also has a positive impact as far as GHG emissions are concerned, since crops and pastureland are responsible for carbon sequestration. We have assumed a carbon sequestration of 110 kg of CO_2 equivalents per stremma[3] for crops (0.3 t C/ha) and 60 kg of CO_2 equivalents per stremma for pastureland (0.16 t C/ha) (see also Pretty and Ball 2001). These sequestration estimations are subtracted from the total emitted GHGs estimated above so that net emissions can be assessed.

15.4 Data

The analysis is undertaken in two sheep farms that represent the extensive and the semi-intensive farming systems and are located in lowland areas of the prefecture of Etoloakarnania, in Western Greece. More specifically, the semi-intensive farm has a flock size of 315 ewes with an annual production of milk about 190 kg/ewe. The live weight of the ewe is 60 kg, and the prolificacy index is 1.5 lambs/ewe. The semi-intensive farm maintains 70 strm of alfalfa and 30 strm of maize for feeding of the flock and utilises 500 strm of pastureland. The milking period is prolonged (from November to July) since there are two lambing periods, in late September and February.

The extensive farm has a flock size of 160 ewes and an annual production of milk of about 100 kg/ewe. The live weight of the ewes and the prolificacy index are also lower in the extensive farm (50 kg/ewe and 1.3 lambs/ewe, respectively). In the farm, 20 strm of alfalfa and 18 strm of maize are cultivated, but the feeding requirements are mainly met through grazing (800 strm of pastureland). Labour inputs are offered mainly by the farmer, and the milking period is smaller than in the case of the semi-intensive farm (January to May). Detailed data from the two farms is used to derive all technical and economic coefficients of the model.

15.5 Application and Results

The mathematical programming model is used to simulate the operation of the two farms, and the optimal farm plan is obtained. This optimal farm plan is used to evaluate the performance of the farms, which is discussed in detail in the following paragraphs.[4] The constraint on net emissions is then inserted, and the optimal farm plan is again obtained for various levels of abatement, through parametric optimisation. This way, the best abatement strategy for each farm can be identified. Finally, the marginal abatement cost for each of the farms is estimated and the marginal abatement cost curve is built.

[3] 1 stremma (strm) = 0.1 ha.

[4] It should be noted that the performance of the mathematical model is satisfactory, since the optimal farm plan is very close to the observed one, especially in the case of the semi-intensive farm.

15.5.1 Socio-economic Performance

Table 15.1 contains the optimal farm plan for the semi-intensive farm. The total gross margin and the gross margin per ewe are 56,775 € and 179 €, respectively. According to Kitsopanides (2006), semi-intensive farms are considered profitable and have an annual net return of 29.4 €/ewe. Although the model used in this analysis maximises gross margin, fixed cost is known and can be used to evaluate net return, which is 45.2 €/ewe, indicating that the economic performance of the semi-intensive farm is very satisfactory. As far as the employment level is concerned, the farm offers full-time employment to the two owners, since family labour is 3,463 h and extra hired labour is also required (87 h).

On the other hand, the extensive farm has a lower gross margin per productive ewe (117 €) (Table 15.2). According to Kitsopanides (2006), extensive farms have a negative net return (–5.6 €/ewe). In this analysis, the net return of the extensive farm is small but positive (6.4 €/ewe), indicating that the activity is viable. Labour inputs per ewe are higher compared to the semi-intensive farm due to the extra labour required for grazing and to the limited invested capital (e.g. absence of milking machine).

The environmental performance of the two farms is discussed in detail in the next paragraph. Tables 15.1 and 15.2, however, contain the optimal farm plan for the farms under the hypothesis of various levels of abatement, or in other words the optimal abatement strategy for the farms. A 10% abatement for the semi-intensive farm leads to a 5% reduction of the gross margin and an 11% reduction of labour (Table 15.1). At a 20% abatement level, the reduction in gross margin and labour is 10% and 22%, respectively, and full-time employment is offered to only one of the owners. The overall reduction of gross margin is 5,729 €, and the average abatement cost is 18 €/ewe, which can be used as an indication of the compensation/ewe the farmer should receive for abating.

In the case of the extensive farm, 10% abatement causes a trivial (less than 1%) reduction of the gross margin (Table 15.2). The reason for this is that the gross margin generated by the sheep farming activity is, in the case of the extensive farm, very small compared to the semi-intensive farm, mainly due to low productivity. Thus, the income loss can easily be replaced by the income generated from cash crop production. The substitution of the sheep farming activity with cash crops is evident in both production systems. The difference is that only in the case of the extensive farming system can this substitution compensate for the income lost from the restriction of the sheep farming activity. Table 15.2 also indicates that 20% abatement causes only a 5% reduction on the gross margin of the farm. On the other hand, abatement has a significant impact on the employment level of the extensive farm as well, since sheep farming, which has high labour requirements, is gradually abandoned. Specifically, 10% and 20% abatement cause 10% and 17% reduction in labour, respectively.

Table 15.1 Optimal solution of the semi-intensive farm for different abatement levels

	0		0.10		0.15		0.20	
Abatement (α)	Total	Per ewe	Total	Per ewe	Total	Per ewe	Total	Per ewe
Gross margin (€)	56,775	179	53,851	190	52,450	198	51,046	207
Total labour (hour)	3,550	11	3,161	11	2,960	11	2,760	11
Productive ewes	318	1	283	1	265	1	247	1
Produced maize for consumption (kg)	47,060	148	30,876	109	22,341	84	13,740	56
Purchased maize (kg)	16,638	52	24,296	86	28,300	107	32,324	131
Produced alfalfa for consumption (kg)	90,356	284	76,887	272	69,620	263	62,455	253
Purchased alfalfa (kg)	0	0	0	0	0	0	0	0
Grass consumed (kg)	250,000	786	246,529	871	247,072	932	247,574	1,002
Produced maize for sale (kg)	0	–	0	–	0	–	0	–
Produced alfalfa for sale (kg)	820	–	34,517	–	52,454	–	70,369	–

Table 15.2 Optimal solution of the extensive farm for different abatement levels

Abatement (α)	0		0.10		0.15		0.20	
	Total	Per ewe	Total	Per ewe	Total	Per ewe	Total	Per ewe
Gross margin (€)	19,952	117	19,868	130	19,568	132	19,022	137
Total labour (hour)	2,510	15	2,272	15	2,203	15	2,080	15
Productive ewes	171	1	153	1	148	1	139	1
Produced maize for consumption (kg)	20,538	120	13,917	91	12,987	88	11,328	81
Purchased maize (kg)	0	0	0	0	0	0	0	0
Produced alfalfa for consumption (kg)	14,753	86	7,853	51	16,480	111	15,478	111
Purchased alfalfa (kg)	4,454	26	9,183	60	0	0	0	0
Grass consumed (kg)	320,000	1,871	320,000	2,092	313,073	2,115	302,088	2,173
Produced maize for sale (kg)	0	–	0	–	0	–	0	–
Produced alfalfa for sale (kg)	8,298	–	23,079	–	15,559	–	18,537	–

15.5.2 Environmental Performance

The environmental performance of the semi-intensive farm and the extensive farm and specifically their GHG emissions are presented in Tables 15.3 and 15.4. Specifically, Tables 15.3 and 15.4 contain the overall emissions of the farms as well as the emissions per main source.

The net emissions (total emissions minus carbon sequestration) are also presented. The main source of GHGs in sheep farms is enteric fermentation, since it is responsible for 71% of the total emitted GHGs of the semi-intensive farm and 78% of the total emitted GHGs of the extensive farm. Similar findings on the contribution of CH_4 emissions in ruminant livestock farms have been reported in previous studies (e.g. Smith and Upadhyay 2005; Petersen et al. 2009). Twenty-one per cent of the emissions of the semi-intensive farm refer to N_2O emissions, and the remaining 8% are CO_2 emissions (Table 15.3). As far as the extensive farm is concerned, N_2O is responsible for 16% of the total emitted GHGs, and CO_2 accounts for only 6% of the total emitted GHGs (Table 15.4). Emissions from enteric fermentation per kg of milk are higher in the case of the extensive farm because of the higher participation of primarily grass and secondarily alfalfa in livestock feeding.

Net emissions of the semi-intensive farm are about 180 t or 2.99 kg of CO_2 equivalents/kg of milk. For the extensive farm, net emissions are significantly higher, reaching 5.45 kg of CO_2 equivalents/kg of milk (net emissions are over 93 t). Tables 15.3 and 15.4 also contain emissions by source, at various levels of abatement. Abatement is in both production systems accompanied by a switch towards cash crops production and, specifically, alfalfa production. This is because alfalfa requires less fertiliser inputs and therefore causes fewer emissions. Furthermore, alfalfa generates a high gross margin, since it has a high yield. The substitution of the sheep farming activity with crop activities has also been pointed out in the study of Petersen et al. (2009) on GHG abatement in extensive grazing systems of south-western Australia. As can be observed in Table 15.3, abatement in semi-intensive farms is achieved by reducing CH_4 emissions from enteric fermentation and N_2O emissions from livestock and maize production.

Specifically, in order to achieve a 10% abatement of net emissions in the semi-intensive farm, CH_4 emissions are reduced by 9%. This reduction is achieved by the reduction of the number of ewes by 35. The analysis indicates, however, that the semi-intensive farms continue to utilise their pastureland, even though grass consumption causes CH_4 emissions. This is probably because in our analysis, carbon sequestration of pastureland is also taken under consideration. Similarly, a 20% reduction of net emissions leads to an 18% reduction of CH_4 and a 45% reduction in N_2O emissions from crops. Abatement is again achieved through the reduction of the number of ewes and a switch towards cash crops. It should also be mentioned that as the level of abatement increases, semi-intensive farms rely more on concentrates (maize) for the feeding of the flock, since less alfalfa is used in the ration.

In the case of the extensive farm, abatement is again achieved through change in production orientation from sheep to cash crops (Table 15.4). As mentioned above,

Table 15.3 GHG emissions in kg of CO_2 equivalents of the semi-intensive sheep farm for different abatement levels

	0		0.10		0.15		0.20	
Abatement (α)	Total	Per kg of milk	Total	Per kg of milk	Total	Per kg of milk	Total	Per kg of milk
Net emissions	180,452	2.99	162,407	3.02	153,384	3.05	144,362	3.08
Total GHGs	211,466	3.50	194,187	3.61	185,564	3.69	176,944	3.77
CH_4 emissions	149,545	2.48	135,844	2.53	129,066	2.56	122,290	2.61
N_2O excreta	37,987	0.63	33,806	0.63	31,655	0.63	29,505	0.63
N_2O fertiliser	4,129	0.07	3,221	0.06	2,743	0.05	2,260	0.05
CO_2 energy	14,714	0.24	13,881	0.26	13,441	0.27	12,998	0.28
N_2O fertiliser – purchased feed	1,814	0.03	2,648	0.05	3,085	0.06	3,523	0.08
CO_2 energy – purchased feed	3,278	0.05	4,786	0.09	5,575	0.11	6,368	0.14

Table 15.4 GHG emissions in kg of CO_2 equivalents of the extensive sheep farm for different abatement levels

	0		0.10		0.15		0.20	
Abatement (α)	Total	Per kg of milk	Total	Per kg of milk	Total	Per kg of milk	Total	Per kg of milk
Net emissions	93,174	5.45	83,856	5.48	80,129	5.41	74,539	5.36
Total GHGs	127,070	7.43	118,131	7.72	113,661	7.68	108,079	7.78
CH_4 emissions	99,111	5.80	92,902	6.07	90,310	6.10	86,000	6.19
N_2O excreta	17,026	1.00	15,234	1.00	14,736	1.00	13,840	1.00
N_2O fertiliser	3,187	0.19	2,342	0.15	2,223	0.15	2,012	0.14
CO_2 energy	7,184	0.42	6,496	0.42	6,400	0.43	6,227	0.45
N_2O fertiliser – purchased feed	102	0.01	211	0.01	0	0.00	0	0.00
CO_2 energy – purchased feed	459	0.03	946	0.06	0	0.00	0	0.00

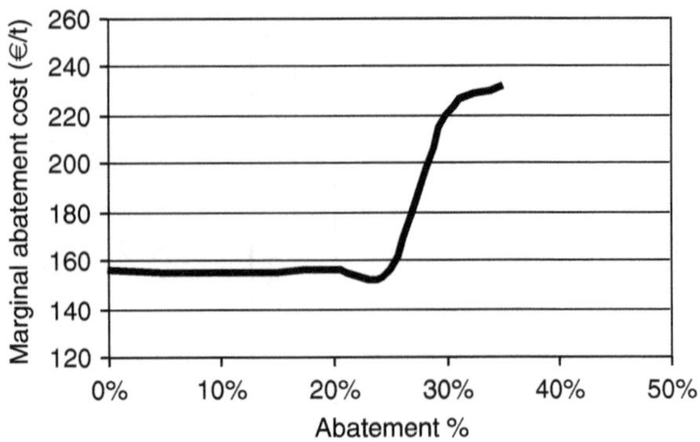

Fig. 15.1 Marginal abatement cost of the semi-intensive farm for different abatement levels

this substitution minimises the impact of abatement in extensive farms, due to the low gross margin of sheep farming and the high yield of alfalfa. In the case of the extensive farm, CH_4 emissions from enteric fermentation are reduced only by 6% and 13% in order to achieve 10% and 20% abatement, respectively. This reduction of CH_4 emissions is achieved only by reducing the flock size and not by increasing the proportion of concentrates used in the ration.

15.5.3 Abatement Cost

Figure 15.1 presents the abatement cost curve for the semi-intensive farm. As can be seen in this figure, the curve indicates an increasing marginal abatement cost. This marginal abatement cost is 156 €/t until 25% abatement is achieved. Then the abatement cost increases to reach 220 €/t and 232 €/t, at 30% and 35% abatement, respectively. These findings denote the significant impact that abatement has on the gross margin of semi-intensive farms, since the income generated by crop production cannot replace the income lost from the restriction of the sheep farming activity.

The abatement cost curve of the extensive farm is presented in Fig. 15.2. As in the case of the semi-intensive farm, the marginal abatement cost of the extensive farm is also increasing, with an average of 50 €/t, until a 20% abatement is reached. The shadow price of net emissions is very small at current emission levels (5 €/t) and gradually increases to 10 €/t at 10% abatement, 91 €/t at 20% abatement, 154 €/t at 25% abatement and 218 €/t at 50% abatement.

Breen and Donnellan (2009) estimated a marginal abatement cost of 110–230 €/t for dairy farms in Ireland, while De Cara and Jayet (2000) estimated the marginal cost, that varied significantly amongst farm types, from 30 to 300 €/t. The low abatement cost of the extensive farm, until 25% abatement is achieved, is explained by the substitution of sheep farming with almost equally profitable crop activities. These results support the heterogeneity of the marginal abatement cost within the sheep farming activity in Greece. The heterogeneity of the GHG abatement cost has been pointed out in a number of studies (e.g. De Cara and Jayet 2000).

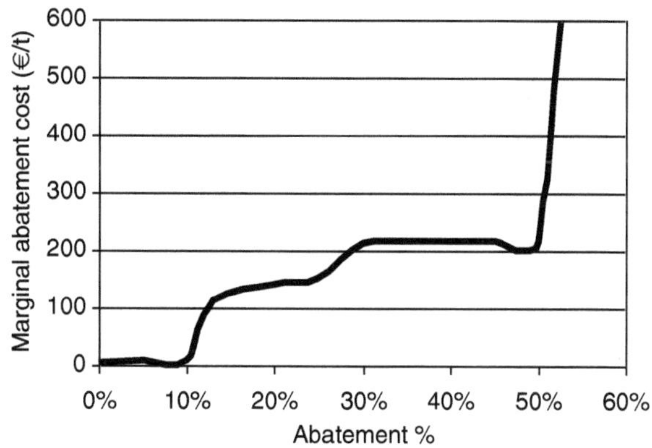

Fig. 15.2 Marginal abatement cost of the extensive farm for different abatement levels

Assessing the marginal abatement cost is useful to policymakers who wish to develop well-targeted and well-designed abatement policy measures. One potential policy measure is the implementation of a tax per ton of emitted CO_2 equivalents (Neufeldt and Schäfer 2008; Petersen et al. 2009). The analysis can assist in the determination of the level of this tax according to the abatement cost of the farms (see also De Cara and Jayet 2000). If a tax smaller than the marginal abatement cost of a farm is implemented, then the farmer will choose to pay the implemented tax instead of abating, and, thus, the policy measure will be ineffective. For example, according to our analysis, a tax of 90 €/t of CO_2 equivalents will have no effect on the emissions of the semi-intensive farms but will succeed to reduce emissions of the extensive farm up to 20%. Furthermore, this tax will also have an impact on the sustainability of the extensive farming system since it can lead to the restriction of the sheep farming activity.

15.6 Concluding Remarks

In this study, a mathematical programming model was used to derive the optimal farm plan of sheep farms and estimate their socio-economic and environmental (in terms of GHG emissions) performance. The abatement strategy and the marginal abatement cost of sheep farms are also estimated. The analysis is undertaken in two sheep farms that represent the semi-intensive and the extensive production systems and includes pre-chain emissions as well as all potential emission sources in the farm. The model maximises gross margin that is used as an economic sustainability indicator. Labour inputs are used as a social performance indicator and GHG emissions as an environmental sustainability indicator.

The results of the analysis indicate that both production systems are economically viable, though the semi-intensive farm has a higher gross margin than the extensive one. The main source of GHG emissions in dairy sheep farms is enteric fermentation. Emissions are particularly high in extensive farms, because of the excessive use of grass and alfalfa for feed. Thus, the semi-intensive system is more efficient in socio-economic but also environmental terms. Across various abatement levels, the optimal solution indicates that abatement is achieved in both production systems, via a switch to cash crops. This has a significant impact only on the gross margin of the semi-intensive farms that are characterised by high productivity. On the other hand, abatement has a significant impact on employment, since the labour requirements of crop production are significantly lower than the labour requirements of sheep farming. As far as the marginal abatement cost is concerned, it is increasing across various levels of abatement, and it is significantly higher in the case of the semi-intensive farm.

The results of the analysis of the two farms are an indication of the heterogeneity of the abatement cost amongst sheep farms with different characteristics. Utilising a farm typology can reflect this heterogeneity more accurately and can be used to estimate the total cost of abating for the country. However, the results of the analysis have highlighted some aspects of the sustainability of the sheep farming activity and can be used as a guide for the development of effective mitigation policy measures.

Acknowledgements This research has been co-financed by the European Union (European Social Fund – ESF) and Greek national funds through the Operational Programme "Education and Lifelong Learning" of the National Strategic Reference Framework (NSRF) – Research Funding Program: Heracleitus II – investing in knowledge society through the European Social Fund.

References

Alford, A. R., Griffith, G. R., Cacho, O. (2004, March). *A Northern Tablelands whole-farm linear program for economic evaluation of new technologies at the farm-level.* (Economic Research Report No. 13), Retrieved from NSW Website: http://www.dpi.nsw.gov.au/research/areas/biosecurity/economics-research/reports/err13

Audsley, E., Stacey, K., Parson, D. J., Williams, A. G. (2009). *Estimation of the greenhouse gas emissions from agricultural pesticide manufacture and use*. Retrieved December 1, 2009, from Cranfield University. Prepared for Crop Protection Association. Web site: https://dspace.lib.cranfield.ac.uk/bitstream/1826/3913/1/Estimation_of_the_greenhouse_gas_emissions_from_agricultural_pesticide_manufacture_and_use-2009.pdf

Benoit, M., & Laignel, G. (2008, July). *Sheep for meat farming systems in French semi-upland area. Adapting to new context: increased concentrates and energy prices, and new agricultural policy* (Paper presented at the 8th Symposium of the International Farming Systems, Clermont-Ferrand, France).

Breen, J., Donnellan, T. (2009, March–April). *Estimating the marginal cost of greenhouse gas emissions abatement using Irish farm-level data.* (Paper presented at the 83rd Annual Conference of the Agricultural Economics Society, Dublin, Ireland).

Crosson, P., O' Kiely, P., O' Mara, F. P., & Wallace, M. (2006). The development of a mathematical model to investigate Irish beef production systems. *Agricultural System, 89*, 349–370.

De Cara, S., & Jayet, P. A. (2000). Emissions of greenhouse gases from agriculture: The heterogeneity of abatement costs in France. *European Review of Agricultural Economics, 27*(3), 281–303.

H.M.R.D.F. (2007). *Sheep and Goat Sector Development*. Retrieved October 17, 2007 from: http://www.minagric.gr/greek/ENHM_FYLADIA_ZWIKHS/ENHM_FYL_zoiki.htm (in Greek).

IPCC. (2006). *IPCC Guidelines for national greenhouse gas inventories*. Retrieved March 6, 2010 from Intergovernmental Panel on Climate Change: Bracknell, U.K. Web site: http://www.ipcc-nggip.iges.or.jp/public/2006gl/vol4.html

Kalaisakis, P. (1965). *Applied animal nutrition*. Athens: Giourdas M. Publishing. in Greek.

Kitsopanides, G. (2006). *Economics of animal production*. Thessaloniki: ZITI Publishing. in Greek.

Neufeldt, H., & Schäfer, M. (2008). Mitigation strategies for greenhouse gas emissions from agriculture using a regional economic-ecosystem model. *Agriculture, Ecosystems & Environment, 123*, 305–314.

Olesen, J. E., Schelde, K., Weiske, A., Weisbjerg, M. R., Asman, W. A. H., & Djurhuus, J. (2006). Modelling greenhouse gas emissions from European conventional and organic dairy farms. *Agriculture, Ecosystems & Environment, 112*, 207–220.

Petersen, E. H., Schilizzi, S., Bennet, D. (2009). *The impacts of greenhouse gas abatement policies on the predominantly grazing systems of south-western Australia*. Retrieved October 2, 2009 from the Australian National University, Asia Pacific School of Economics and Government, International and development Economics: Working Paper 02–9 Web site: http://www.crawford.anu.edu.au/degrees/idec/working_papers/IDEC02-9.pdf

Pitesky, M., Stachhouse, K. R., & Mitloehner, F. M. (2009). Clearing the air: Livestock's contribution to climate change. *Advances in Agronomy, 103*, 1–40.

Pretty, J., & Ball, A. (2001). *Agricultural influences on carbon sequestration: A review of evidence and the emerging trading options*. Retrieved October 2, 2009, from the University of Essex, Center for Environment and Society and Department of Biological Studies, Occasional Paper 2001–2003, Website: http://citeseerx.ist.psu.edu/viewdoc/download?doi=10.1.1.188.7509&rep=rep1&type=pdf

Schils, R. L. M., de Haan, M. H. A., Hemmer, J. G. A., van den Pol-van Dasselaar, A., de Boer, J. A., Evers, A. G., Holshof, G., van Middelkoop, J. C., & Zom, R. L. G. (2007). DairyWise, a whole-farm dairy model. *American Dairy Science Association, 90*, 5334–5346.

Smith, E. G., & Upadhyay, M. (2005, July). *Greenhouse gas mitigation on diversified farms*. (Paper presented at the Western Agricultural economics Association-Western Economics Association International Joint Annual Meeting, San Francisco, CA, USA).

Vermorel, M., Jouany, J.-P., Eugene, M., Sauvant, D., Noblet, J., & Dourmad, J.-Y. (2008). Evaluation quantitative des émissions de méthane entérique par les animaux d'élevage en 2007 en France. *INRA Productions Animales, 21*(5), 403–418.

Veysset, P., Bedin, D., & Lherm, M. (2005). Adaptation to Agenda 2000 (CAP reform) and optimization of the farming system of French suckler cattle farms in the Charolais area: A model-based study. *Agricultural Systems, 83*, 179–202.

Veysset, P., Lherm, M., & Bédin, D. (2010). Energy consumption, greenhouse gas emissions and economic performance assessments in French Charolais suckler cattle farms: Model-based analysis and forecasts. *Agricultural Systems, 103*, 41–50.

Weiske, A., Vabitsch, A., Olesen, J. E., Schelde, K., Michel, J., Friedrich, R., & Kaltschmitt, M. (2006). Mitigation of greenhouse gas emissions in European conventional and organic dairy farming. *Agriculture, Ecosystems & Environment, 112*, 221–232.

Wood, S., & Cowie, A. (2004, July). A review of greenhouse gas emission factors for fertilizer production. *IEA Bioenergy Task 38*. Retrieved from http://www.ieabioenergy-task38.org/publications/GHG_Emission_Fertilizer%20Production_July2004.pdf

Zervas, G., Kalaisakis, P., & Feggeros, K. (2000). *Farm animal nutrition*. Athens: Stamoulis Publishing. n Greek.

Part VI
Sustainability Assessment in Non-European Farming Systems

Chapter 16
Method for the Evaluation of Farm Sustainability in Quebec, Canada: The Social Aspect

Diane Parent, Valérie Bélanger, Anne Vanasse, Guy Allard, and Doris Pellerin

Abstract In Quebec, over the last two decades, the number of dairy farms has declined, and the sustainability of the family farms and their rural communities has been questioned. In order to be sustainable, a farm should be viable, liveable, transmissible and ecologically reproducible. Thus, the assessment of farm sustainability should be based on its economic, environmental and social aspects. A holistic method, named DELTA, was developed for the three aspects: environmental, technical-economic and social. To identify the indicators, we used a multiple stakeholder perspective (researchers, farmers, advisors). We report on the 20 social indicators that were selected and grouped in four components: quality of life, social integration, farm succession and entrepreneurial skills. Results are presented as a radar diagram for each farm with axis representing indicators. Each indicator was validated on 40 farms. This new assessment tool will serve to evaluate the sustainability of a farm at a given point in time and could be used periodically to follow its evolution over the years.

Keywords Farm sustainability • DELTA method • Dairy farms • Social indicators • Sustainability assessment

D. Parent (✉) • D. Pellerin
Département des sciences animales, Faculté des sciences de l'agriculture et de l'alimentation, Université Laval, Québec, Province de Québec, Canada G1V 0A6
e-mail: diane.parent@fsaa.ulaval.ca; doris.pellerin@fsaa.ulaval.ca

V. Bélanger • A. Vanasse • G. Allard
Département de phytologie, Faculté des sciences de l'agriculture et de l'alimentation, Département des sciences animales, Université Laval, Québec, Province de Québec, Canada G1V 0A6
e-mail: valerie.belanger@fsaa.ulaval.ca; anne.vanasse@fsaa.ulaval.ca; guy.allard@fsaa.ulaval.ca

A.A. Marta-Costa and E. Silva (eds.), *Methods and Procedures for Building Sustainable Farming Systems: Application in the European Context*,
DOI 10.1007/978-94-007-5003-6_16,

16.1 Introduction

It seems that the concept of sustainable development (SD) is becoming widespread. Agriculture is no exception, and agricultural systems will have to evolve in this direction. According to the FAO (1991), a sustainable agricultural system is a system that preserves land and water resources, that preserves genetic resources of plants and animals and that is both economically viable and socially acceptable. But how can we assess the level of conservation of resources as well as the economic viability and social acceptability of an agricultural system?

In Quebec (Canada), several factors lead us to take a better look at the sustainability of our farms. The intensification of production and geographic concentration of certain crops and livestock have resulted in an increased pressure on water, air and soil resources. Also, certain economic conditions and an ongoing uncertainty about the future of milk quotas undermine the economic situation of Quebec farms. The decline of the number of dairy farms in Quebec is also a major concern. In addition, societal expectations are increasingly high towards the modes of production and towards the quality of rural areas in general. Therefore, the concept of sustainability has also become a challenge for social sciences. Many actors advocate greener farms. But, will there even be enough farmers to manage those farms? Will the farmers continue to like their work? The rate of farmer generation renewal continues to decline, going from 53% in 2001 to 35% in 2006 (Stat Can 2007). Consequently, defining what a sustainable agricultural system has become a real necessity.

In order to meet this challenge, our research project's objective is to develop a method to assess the overall sustainability of dairy farms that integrates the three aspects of SD. The method, called *DELTA*, is a farm self-diagnostic tool that can follow a farm's progress on its path to sustainability by tracking its adoption of best practices. It was named *DELTA* because this word often represents a triangle at the confluence of several rivers. The purpose of this chapter is to illustrate the approach that was used for the construction of this assessment tool, specifically the social aspect.

16.2 An All-Encompassing Definition of Sustainable Farming

The concept of sustainable agriculture boasts fairly extensive literature, but its application on the farm raises several interpretations. Few sustainability assessment tools or methods that include all three aspects of SD are available at the farm level. For the majority of people and organizations, SD was first and foremost an environmental concept, and only this aspect deserved assessment (Häni 2006). This was a misconception since sustainability is defined as environmentally liveable, economically viable and socially equitable.

Even though it has been several years since SD was described, defining the sustainable farm concept is different. Our definition implies that the farm must be viable, liveable, transmissible and ecologically reproducible (Landais 1998; Parent 2001).

"Reproducibility" is defined by the logical use of natural resources (water, soil, air) through good agricultural practices and the reproducible potential of those resources. "Viability" depends on the good technical-economic performance of the farm and implies that the farm must be able to generate secure long-term income. "Liveability" is a concept that reflects the quality of life of farmers and their families, both on the farm and in their community. Finally, "transmissibility" expresses the farm's potential to be taken over through succession by the next generation as well as the role of agriculture in the dynamics of local development. This definition is a basic premise when the objective is to assess sustainability at the farm level.

16.3 Methods for the Assessment of Sustainability

Although literature is abundant on SD indicators and on the assessment of sustainability in general, few documents on overall sustainability include the three aspects. However, there are some tools that take into account environmental, economic and social indicators. Some conceptual approaches extend the assessment of sustainability to the social aspect, but there are very few of those approaches, and they are harder to discern (von Wiren-Lehr 2001; Rasul and Thapa 2004).

In the Netherlands, a study by Van Calker et al. (2005) presents the sustainability of dairy farms according to four aspects: economic, ecological, internal social and external social. In total, 12 components were assigned to the ecological and external social aspects, three to the economic aspect and only one to the internal social aspect (working conditions), while five indicators measure the external social aspect (food safety, animal welfare, animal health, landscape quality and cattle grazing).

In France, several methods have been developed to assess farms from all angles. One particular method, which has been tested on many farms, is presented below. It is called the IDEA method which is an acronym for I*ndicateurs de* d*urabilité des* e*xploitations* a*gricoles* (Vilain 2001). It assigns scores to the farmer's practices and behaviour and measures the sustainability of different types of farms in France by using three scales: the agroecological, socio-territorial and economic scales. These three scales of sustainability contain components, which in turn contain indicators. These indicators represent the variables to be assessed. There are 19 indicators for the agroecological scale, six for the economic scale and 16 for the socio-territorial scale. These three scales of sustainability have the same weight.

Closer to home, researchers at the University of Guelph in Ontario, Canada, in partnership with Switzerland, have developed a tool for assessing the overall sustainability of farms (Häni et al. 2003). Their tool is called RISE, which stands for *R*esponse-*I*nducing *S*ustainability *E*valuation. Table 16.1 summarizes the strengths and weaknesses of the existing methods that were just reviewed.

Although far from exhaustive, this overview of literature nevertheless gives us an idea of the indicators that play a role in the assessment of each dimension of SD. It is vital for Quebec to develop its own suitable and adapted method.

Table 16.1 Summary of the strengths and weaknesses of the existing methods presented

Method	Origin	Strength	Weaknesses
Van Calker et al.	Netherlands	Takes into account the 3 aspects	Only one economic indicator
		Touches the social aspect	No indicator on the quality of life of families
		The indicators are specific to the dairy sector	
Vilain et al. (IDEA)	France	Allows a constant evolution of the method	Indicators are not specific to one production
		Takes into account the 3 aspects	
		The indicators are related to sustainability goals	The calculation of scores requires a lot of data
		The indicators are grouped into components	
Häni et al. (RISE)	Switzerland and Ontario	Small number of indicators (12) to cover the 3 aspects	Few social indicators
	(Canada)	State and pressure on the resource for each indicator	Indicators are not specific to one production

16.4 Why Use Indicators?

The indicator concept is not new. In the past few years, the social and economic domains have been using several trusted indicators that have proven to be valuable governance tools. Examples include the gross national product or the civil status of citizens. Indicators of SD have also been recently integrated into agriculture. Extensive literature exists about these indicators, but since they were developed at the political level, they are often only applicable at the national or international levels. Not a lot of work has been done on the development of indicators that are applicable at the farm level.

Indicators are variables that provide information on other variables that are more difficult to access. They can be used as reference points to make decisions (Gras et al. 1989 cited by Bockstaller and Girardin 2003; van der Werf and Petit 2002). An indicator is a tool that reduces the complexity of descriptions and integrates information from a system (Giampietro 1997 cited by von Wiren-Lehr 2001). Therefore, the selected indicators must reflect the hard-to-collect information of the farm. For example, soil erosion, which is difficult to measure on the farm without expensive measuring tools, can be indirectly determined through related variables such as soil cover in the fall or the presence of riparian strips, etc.

Certain criteria are essential in choosing the right indicators in order to identify if some indicators are not suitable for a farm-level assessment tool. Table 16.2

Table 16.2 Questions asked to evaluate each indicator in regard to each criterion

Practicality	Easy to implement	Is the data for this indicator already being measured for another purpose?
		Is the data easily available on the farm?
	Comprehensible immediately	Does this indicator need a level of technical expertise to be comprehensible?
	Reproducible (reliability)	Will this indicator be calculated the same way on each farm, year after year?
Usefulness	Sensitive to variations	Will this indicator be able to follow the evolution on the farm, and will the result increase or decrease depending on the management practices adopted by the farmer?
	Adapted to the objectives	Does this indicator answer the objectives "self-diagnosis" and "decision aid tool"?
		Is this indicator linked with environmental sustainability objectives overall?
	Relevant for users	As a primary goal, is this indicator useful for farmers or for other groups as well (advisors, stakeholders, etc.)?

Adapted from Girardin et al. (1999), Hagan and Whitman (2006)

shows these criteria and with which questions they are related to. A distinction is made between criteria used to evaluate the usefulness and criteria used for practicality.

16.5 The Development of the *DELTA* Method: A Social Construction of Indicators

Although the scope of this method seems promising for farmers and the surrounding environment, its interest also lies in its methodological approach which is based on a multidisciplinary approach and the participation of local actors. The Delphi technique (described below) and focus groups were used to consult with experts (researchers, farmers and stakeholders) in order to build and select the indicators. Creating a tool requires a knowledge from the base, i.e. from the field. It also requires an ongoing exchange between the researchers and the people in the field. This approach in the research world is called the inductive exploratory approach (Lessard-Hébert et al. 1996). In the field of agriculture, researchers more frequently use the hypothetico-deductive approach, which consists of making hypotheses and then confirming or disproving them (Van der Maren cited by Lessard-Hébert et al. 1996). This epistemological reminder allows us to better understand how the *DELTA* method was developed with the help of other techniques described below.

16.5.1 The Delphi Technique

The Delphi technique was used here to identify the needed elements for the construction of the indicators for the three aspects of sustainability. It is important to mention that the Delphi technique is done remotely. First, a form containing a question regarding the research subject was sent to the selected experts. In our case, the respondents were asked to list elements to be considered when assessing the sustainability of dairy farms. Potential indicators were compiled and submitted to the same 25 experts who rated them for their relevance and easiness of on-farm acquisition (van der Werf and Petit 2002). Thus, the participants, or experts, have the opportunity to make their points of view evolve. The main features of the Delphi technique are its anonymity, which aims to reduce the influence of "superexperts" and controlled feedback (Mayer and Ouellet 2000).

The downside of this method is that it does not take into account the evolution of points of view that often arises in a group (Clément and Madec 2006). Also according to Rigby et al. (2001), using consultation in order to define the indicators and their weighting can help build a consensus between actors who have different technical opinions. In our case, this recommendation has been included in our methodological process. The focus groups that followed the Delphi enabled us to benefit from these group dynamics.

16.5.2 Focus Groups

Three focus groups were conducted, each with different experts, depending on the aspect. Focus groups are specific to the inductive approach and allow collecting new information that would otherwise be less accessible (Mayer and Ouellet 2000). The focus group members were selected among the participants of the previous Delphi technique. Threshold values were determined for each selected indicator. The indicators were also grouped by component, and their weighting discussed by the group members. For example, the social aspect contains a component called *farm succession* that includes the following four indicators: continuity value, presence of farm succession, preparation for retirement and farm succession integration. Indicators are grouped this way for all three sustainability aspects.

The selected experts all had very different backgrounds. For the social aspect, the experts consisted of entrepreneurship researchers, farmers, farm transfer consultants from the regional farm succession centres and psychologists working mostly in the agricultural community coming to talk about quality of life. Table 16.3 illustrates our project's methodology by summarizing the experts' participation at each stage of the project and by giving the number of statements that have been put forward.

As shown in the last line of Table 16.3, the tool now contains 13 environmental indicators, 8 technical-economic indicators and 20 social indicators. Table 16.4 lists

Table 16.3 Series of consecutive steps of participatory processes

	Environmental aspect		Technical-economic aspect		Social aspect	
Steps	Number of experts	Number of statements kept at the end	Number of experts	Number of statements kept at the end	Number of experts	Number of statements kept at the end
Initial sending	29	0	19	0	32	0
1st Delphi	25	102	19	139	27	166
2nd Delphi	20	23	19	20	18	30
Focus group	12	13	10	8	11	20

Table 16.4 Four components and 20 indicators for the social aspect of the *DELTA* method

Aspects	Components	Indicators
Social sustainability	Quality of life (25 %)	Work and workload, holidays, satisfaction, social support, health and stress, social and professional relationships
	Social integration (15 %)	Contribution in local services, agricultural neighbourhood, quality of non-agricultural relationships, social contribution, regional presence of agriculture
	Farm succession (30 %)	Continuity value, presence of farm succession, preparation for retirement, farm succession integration
	Entrepreneurship (30 %)	Formation, use of advisory services, vision, human resources management, entrepreneurial abilities

the components and indicators for the social aspect of sustainability, more specifically addressed here. The final choice of indicators was first achieved by the experts who were attending the focus groups.

Points were awarded by the experts to each component and each indicator in different ways depending on the aspect of sustainability; at the social level, the weight of each component determined by the expert is quality of life (25%), social integration (15%), farm succession (30%) and entrepreneurship (30%). Moreover, each indicator has its own score (see Table 16.5). Several social indicators are qualitative, and in these cases, Likert scales are used to measure perception. For each scale, a weighting score is attributed. An example is presented below in Sect. 16.5. Finally, the results are added up for each indicator, which gives a final score of 100 for social sustainability.

Table 16.5 Results of the 20 social indicators for both regions

Components	Indicators	Maximum score	Bas-St-Laurent Mean	Monteregie Mean
Quality of life (25)	Work and workload	2	0.7	0.6
	Holidays	2	0.9	1.0
	Satisfaction	7	4.1	4.4
	Social support	4	3.3	3.0
	Health and stress	5	1.8	2.1
	Social and professional relationships	5	3.8	3.6
Social integration (15)	Contribution in local services	2	1.5	1.0
	Quality of non-agricultural relationships	6	4.5	3.8
	Agricultural neighbourhood	4	3.5	3.8
	Social contribution	1	1.0	0.8
	Regional presence of agriculture	2	0.8	0.9
Farm succession (30)	Continuity value	9	7.4	8.4
	Presence of farm succession	8	6.6	6.9
	Preparation for retirement	7	3.1	4.1
	Farm succession integration	6	2.3	1.9
Entrepreneurship (30)	Formation	6	3.8	4.8
	Use of advisory services	6	5.6	5.4
	Vision	6	3.6	3.9
	Human resources management	6	4.3	4.7
	Entrepreneurial abilities	6	4.2	4.8
	Total (score)	100	66.3	69.7

16.5.3 The Studied Farm Sample

Once determined, the indicators were tested on 40 farms in two contrasting regions of Quebec (20 farms in each region) referring to their different pedo-climatic conditions, social situations and contexts of production. The Bas-St-Laurent region (BSL) is a more isolated area than the Monteregie (M) region which lies near major urban centres. Even if the farms are in two different regions, general characteristics like herd size and land area are similar (herd size average (number of cows): BSL = 53 and M = 58; land area average (number of hectares): BSL = 140 and M = 153). The supply management system in Quebec's dairy industry, as for the rest of the country, can explain those small differences between farms. Although Quebec has an increasingly diversified agriculture, the dairy sector was chosen because it remains the largest sector, with 23 % of farms (Statistics Canada 2007). Moreover, Quebec is the leading province in milk production in Canada (Statistics Canada 2007). The contrasting regions enabled to verify if indicator calculations can be done in different contexts and if threshold values are robust enough. To test the selected social indicators, a questionnaire addressed to the farmers was developed, mailed and completed by the owner-operators of the farms.

16.6 Results for Social Indicators of Sustainability

Table 16.5 presents the results for the social aspect of the sustainability that was evaluated in this project. The results are also presented according to each of the two main studied regions. Although the project's objective is not to compare these two regions, this separation of results allows us to see if the chosen thresholds for each indicator have amplitude or not.

The two regions are similar in terms of results. Some indicators (Work and workload, Farm succession integration and Regional presence of agriculture) have low scores compared to the maximum score for each indicator. The indicators *Satisfaction, Vision* and *Preparation for retirement* are slightly better, but work still needs to be done by Quebec dairy farms in those areas. Social indicators are often qualitative and mostly assess the perception of farmers towards the subject. The following example illustrates the Likert scale and the weighting score for each category of the scale in the three questions asked in order to assess the farmers' satisfaction. The three questions are:

- "How much do you enjoy your regular farm activities?" A lot (3), Enough (1), A little (0) or Not at all (0).
- "What is your level of satisfaction with your life?" Very satisfied (2 points), Rather satisfied (1), Rather unsatisfied (0) or Very unsatisfied (0).
- "How do you rank your overall net household income?" Very satisfied (2), Rather satisfied (1), Rather unsatisfied (0) or Very unsatisfied (0).

Then, the score for *Satisfaction* indicator is the addition of each answer. Consequently, the maximum score of this indicator is seven.

Another example of score attribution is for the *Presence of farm succession* indicator. The maximum score for this indicator is eight. First of all, experts desired to modulate the score in function of farmer's ages for retirement. Then, the score for this indicator varies between four possible answers. If there is a serious and interested farm succession, the score will be eight. If it is clear that there is no farm succession, the score will be zero. Between these two answers, two options are available. If the answer is I do not know if there is farm succession for my enterprise, another question is asked about the time before retirement. If I am at least 5 years from my retirement, the score will be five, and inversely, if I am at less than 5 years of my retirement and still do not have a farm succession, the score will be only two. In summary, scores for this indicator could be zero, two, five or eight points.

The assessment of perception can give reliable results, despite the fact that the indicators are qualitative and therefore more difficult to identify. Another good example is the *Vision* indicator. This indicator is verified by simply asking farmers

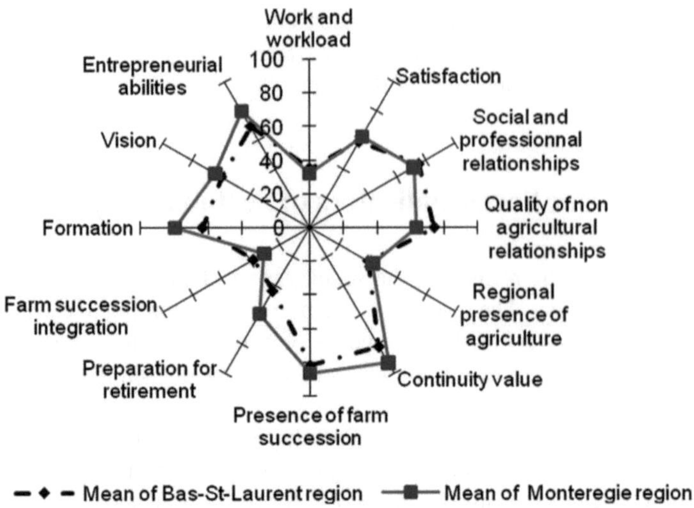

Fig. 16.1 Radar diagram presenting both regions' means for 12 of the 20 social indicators

if they have a clear vision of their farms in 10 years and if they have a written document, or development plan, that expresses orientations, measures and objectives for the future of their farm. These two questions are answered by yes or no. As shown in Table 16.5, for the experts, the most important indicator to assess social sustainability is the *Continuity value* with the highest relative weight of 9 points, followed by *Presence of farm succession* (8) and *Satisfaction* (7).

Results are presented using radar diagrams, also called spider webs (Fig. 16.1). This type of diagram can help to clearly show the strengths and weaknesses of a farm. It contains as many axes as there are indicators. Each axis starts at the centre point and ends at the periphery. The total length of an axis represents the maximum score that can be obtained for the indicator. Straight lines then link the axes, and a polygon is formed. Therefore, the shape of the web varies according to the results. A bigger polygon means a higher score. It is possible for a farm to have a radar diagram that illustrates its results compared to the regional average (Fig. 16.1). Comparing a farm to its region instead of the whole province helps to better guide the farmer concerning his results. This is because different regional contexts can create significant differences. For example, maize cultivation is very present in Monteregie but is rare in the Bas-St-Laurent region. The end result of the diagnostic tool will be these diagrams. The farmers will receive a copy of them after their evaluation. The means for each region will gradually appear as more and more farmers do the evaluation.

Figure 16.1 contains 12 indicators and shows that the two regions have similar trends in terms of social sustainability. The shapes of the two polygons are similar, but Monteregie seems a little stronger when it comes to the entrepreneurship component, especially when it comes to the Vision, Entrepreneurial abilities and Formation indicators. Cohabitation is stronger in the Bas-St-Laurent region.

16.7 Conclusion

In response to the growing popularity of SD, many organizations meet the challenge of building SD indicators for various sectors, including agriculture. We have also chosen to take on this challenge as part of our research project. *DELTA* provides farm-level indicators to assess the overall sustainability of dairy farms. One of the innovative aspects of this tool is that it contains indicators for measuring the social sustainability of dairy farms. The past and future work done with regard to this project clearly demonstrates that the concept of SD is a space where several disciplines meet, confront and collaborate.

A major step that has to be taken is the integration of the three aspects of sustainability. This will provide a new evaluation grid for farms and will illustrate the interrelationships that exist between the components. It will also be useful in order to have a better global and systemic understanding of the farm. This integration will highlight the issues that still need to be clarified concerning SD in agriculture. The grid will serve as a reflection tool for farmers wishing to enter an operational approach to sustainability in Quebec.

Combined with a long-term vision, this method could help improve the effectiveness of environmental and rural policies and help establish the priorities for SD in the agricultural sector. Finally, our understanding of the principles of sustainable agriculture will have improved, and an attempt to operationalize the concept of SD will have been made.

All in all, the creation of a sustainability assessment method will be an asset to the dairy sector in Quebec. In the words of Landais (1998): "The sustainable development discourse is actually a new social contract that is offered to farmers. And we cannot exclude that sustainability will play the same role in the next decade that productivity has played in previous ones". Hopefully, our contribution will allow more farms to survive these next decades.

Acknowledgements The research team wishes to thank the Fonds québécois de recherche sur la nature et les technologies (FQRNT) for the grant and NOVALAIT for the doctoral student's scholarship received by Valérie Bélanger.

References

Bockstaller, C., & Girardin, P. (2003). How to validate environmental indicators. *Agricultural Systems, 76*, 639–653.

Clément, O., & Madec, P. (2006). Un outil pour la construction d'indicateurs de développement durable: la méthode Delphi. *Une expérience en aquaculture. Natures Sciences Sociétés, 14*, 297–302.

FAO/Pays-Bas. (1991). *Stratégies et instruments pour une agriculture et un développement rural durables*, Conférence FAO/Pays-Bas sur l'agriculture et l'environnement, 's-Hertogenbosch, 1991, FAO/Pays Bas.

Girardin, P., Bockstaller, C., & van der Werf, H. (1999). Indicators: Tools to evaluate the environmental impacts of farming systems. *Journal of Sustainable Agriculture, 13*, 5–21.

Hagan, J. M., & Whitman, A. A. (2006 June). Biodiversity indicators for sustainable forestry: Simplifying complexity. *Journal of Forestry, 104*, 203–210.

Häni, J. F. (2006). *Global agriculture in need of sustainability assessment.* Proceedings and outputs of the first Symposium of the International Forum on Assessing Sustainability in Agriculture; Sustainable Agriculture: From Common Principles to Common Practice, Bern, Switzerland, International Institute for Sustainable Development.

Häni, F., Braga, F., Stampfli, A., Keller, T., Fischer, M., & Porsche, H. (2003). RISE, a tool for holistic sustainability assessment at the farm level. *International Food and Agribusiness Management Review, 6*(4), 78–90.

Landais, E. (1998). Agriculture durable: les fondements d'un nouveau contrat social? *Le courrier de l'environnement* no. 33 avril. INRA, Paris, France.

Lessard-Hébert, M., Goyette, G., & Boutin, G. (1996). *La recherche qualitative. Fondements et pratiques* (2nd ed.). Montréal: Éditions Nouvelles.

Mayer, R., & Ouellet F. (2000). *Méthodes de recherche en intervention sociale. Gaëtan Morin éditeurs*. Boucherville, Québec, Canada.

Parent, D. (2001 été). D'une agriculture productiviste en rupture avec le territoire à une agriculture durable complice du milieu rural. *Téoros, 20*(2), 22–25.

Rasul, G., & Thapa, G. B. (2004). Sustainability of ecological and conventional agricultural systems in Bangladesh: An assessment based on environmental, economic and social perspectives. *Agricultural Systems, 79*, 327–351.

Rigby, D., Woodhouse, P., Young, T., & Burton, M. (2001). Constructing a farm level indicator of sustainable agricultural practice. *Ecological Economics, 39*, 463–478.

Statistics Canada, (2007). *Census of Agriculture 2006.* http://www.statcan.gc.ca/cara2006/index-eng.htm. Accessed 15 Nov 2011.

Van Calker, K. J., Berentsen, P. B. M., Giesen, G. W. J., & Huirne, R. B. M. (2005). Identifying and ranking attributes that determine sustainability in Dutch dairy farming. *Agricultures and Human Values, 22*, 53–63.

van der Werf, H. M. G., & Petit, J. (2002). Evaluation of the environmental impact of agriculture at the farm level: A comparison and analysis of 12 indicators-based methods. *Agriculture, Ecosystems & Environment, 1922*, 1–15.

Vilain, L. (2001). La méthode IDEA, indicateurs de durabilité des exploitations agricoles. guide d'utilisation. Educagri Éditions. Dijon, France.

Von Wiren-Lehr, S. (2001). Sustainability in agriculture-an evaluation of principal goal-oriented concepts to close the gap between theory and practice. *Agriculture, Ecosystems & Environment, 84*, 115–129.

Chapter 17
Adapting a European Sustainability Model to a Local Context in Semi-arid Areas of Lebanon

Elias Ghadban, Salma Talhouk, Mabelle Chedid, and Shady K. Hamadeh

Abstract A French agriculture sustainability assessment model (IDEA) was modified to fit the Lebanese agriculture context. IDEA is structured around several objectives grouped together to form three sustainability scales: agro-ecological, socio-territorial and economic scale which in turn are translated into measurable indicators. Based on this model, various components of a farming system are assigned by numerical scores that are then weighted and aggregated to give the farm a score for each of the three scales of sustainability. To fit the Lebanese model, some indicators had to be modified. The modified model was initially tested on three farms, and then a full survey was carried out for 1 year over 34 farmers. The modified IDEA model proved to be a useful assessment tool to guide farmers and development agents in assessing agriculture sustainability of small farms in semi-arid areas. It showed high sensitivity within the Lebanese context unveiling differences between and within farming systems and identifying levels of intervention to improve sustainability.

Keywords Agricultural sustainability • Assessment • IDEA model • Lebanon • Indicators

E. Ghadban
School of Oriental and African Studies, Department of Development studies, University of London, Thornhaugh Street, Russell Square, London WC1H 0XG, UK
e-mail: ghadban_elias@hotmail.com

S. Talhouk
Landscape Design and Ecosystem Management Department, American University of Beirut, Riad El Solh, 1107-2020 Beirut, Lebanon
e-mail: ntsalma@aub.edu.lb

M. Chedid • S.K. Hamadeh (✉)
Environment and Sustainable Development Unit, American University of Beirut, Riad El Solh, 1107-2020 Beirut, Lebanon
e-mail: mabelle.chedid@gmail.com; shamadeh@aub.edu.lb

A.A. Marta-Costa and E. Silva (eds.), *Methods and Procedures for Building Sustainable Farming Systems: Application in the European Context*,
DOI 10.1007/978-94-007-5003-6_17,

17.1 Introduction

The consequences of population increase, globalization and environmental degradation in the past two decades affected agriculture and raised concern about the sustainability of agricultural systems (Salvatore 2004). In Lebanon, agriculture, once a prominent activity, now contributes to less than 10% to the gross national product. Over the past 50 years, the Lebanese agricultural sector has witnessed a major shift from a low-input, extensive farming system aimed at staples and some fruit production to an intensive, land-limited and horticulture-based system limited by major structural and environmental constraints. A significant effort is required to improve agricultural sustainability in Lebanon (Hamadeh et al. 2007). A variety of methods have been proposed for the evaluation of the environmental impacts of farms as sustainability measurement (Von Wiren-Lehr 2001; Halberg et al. 2005; Pacini et al. 2002). One model is the French IDEA (Indicateur de durabilité des exploitations agricoles) sustainability assessment model that was developed by a joint working group between the Directorate General for Teaching and Research (DGER) in the French Ministry of Agriculture and Fisheries and the "Bergerie Nationale" (National Sheep Husbandry Centre) in Rambouillet (Zham et al. 2004). This chapter describes the application of the IDEA model to assess the sustainability of the Lebanese farms in the semi-arid Bekaa region.

17.2 Materials and Methods

17.2.1 Sustainability Assessment Model: IDEA

In order to determine the sustainability of the agricultural systems in the Bekaa plain, the IDEA model was used following minor modifications in indicators. IDEA is structured around 16 objectives (Table 17.1) grouped together to form three sustainability scales: agro-ecological, socio-territorial and economic scale. Each of these three scales is subdivided into three or four components (making a total of 10 components): diversity, organisation of space, farming practices, quality of the products and land, employment and services, ethics and human developments, economic viability, independence, transferability and efficiency. The components are in turn made up of a total 41 indicators. A single objective can contribute to the improvement of several components of sustainability, and indicators are intended to translate these objectives into measurable criteria.

Based on the IDEA model, various components of a farming system are assigned by numerical scores that are then weighted and aggregated to give the farm a score for each of the three scales of sustainability (Zahm et al. 2006). The calculation method is based on a point system with an upper limit. The three sustainability scales are of equal weight and range from 0 to 100 points. All farm information is translated into sustainability units, thus determining the score allocated to each indicator.

Table 17.1 The 16 objectives of the IDEA model

Consistency	Local development
Careful management of nonrenewable natural resources	Quality of life
Preservation and management of biodiversity	Product quality
Soil preservation	Adaptability
Preservation and management of water	Ethics
Citizenship or socially aware practices	Employment
Atmosphere preservation	Landscapes preservation
Human development	Animal well-being

For each of the three sustainability scales, the score consists of the cumulative number of basic sustainability units (or points) awarded to the different indicators in that scale in question, and the higher the score, the more sustainable the farm is considered on that particular scale. Each component has a ceiling value (generally 33 points), thus allowing for a large number of possible combinations leading to the same degree of sustainability. When it comes to the global score aggregating the 3 scales, according to Vilain et al. (2003), the lowest value of the three scales is considered as the final farm sustainability score.

The sensitivity of the IDEA method is such that it can detect differences in sustainability between production systems as well as within the same production system. This sensitivity endows the method with the ability to reveal differences between farms either on the level of the three scales or their components or for a particular indicator (Zahm et al. 2006).

17.2.2 Adaptation of IDEA Indicators to the Lebanese Context

The original model was designed to assess French farms where the average size is 42 ha, and therefore, it was modified in order to fit the Lebanese context, where the average farm size is 1.27 ha (MOE/LEDO 2001) and where 75% of farmers cultivated less than one hectare and (MOA 2007). Accordingly, seven indicators, related to farm size or surface reliance and agricultural income in addition to certain agricultural practices, were modified as follows:

- In the agro-ecological scale: enhancement and conservation of genetic heritage, dimensions of field, organic matter management and fertilization
- In the socio-territorial scale: contribution to employment
- In the economic scale: available income in relation to national legal minimum wage and economic transferability

These indicators were initially tested on 6 farms and used in this study to assess the sustainability of selected organic and conventional farms in the Bekaa valley.

17.2.3 Data Collection

Data for the assessment was gathered through a field questionnaire that was devised to include parameters of the assessment method and related questions.

Thirty-four farms were selected, of which 17 were certified organic farms. The remaining 17 farms were conventional and were chosen based on key criteria, namely, proximity to selected organic farms and similarity in size of useful agriculture surface (UAS) and in cultures. The data collected through the questionnaire and the modifications made to the model were initially tested for this mission on three farms.

17.3 Results and Discussion

17.3.1 Applicability of the IDEA Model to the Lebanese Context

17.3.1.1 System Variation

The IDEA model was able to differentiate dimension score contributing to sustainability. The limiting contributor in both systems (organic and conventional) was the agro-ecological one (Fig. 17.1) with 76% of the studied farms having a deficit in such dimension.

17.3.1.2 Components Variation

Variations between sustainability components were also dissected by the IDEA model showing differences within and across production systems. As shown in Fig. 17.2, all the components of agro-ecological (diversity, farming practices and

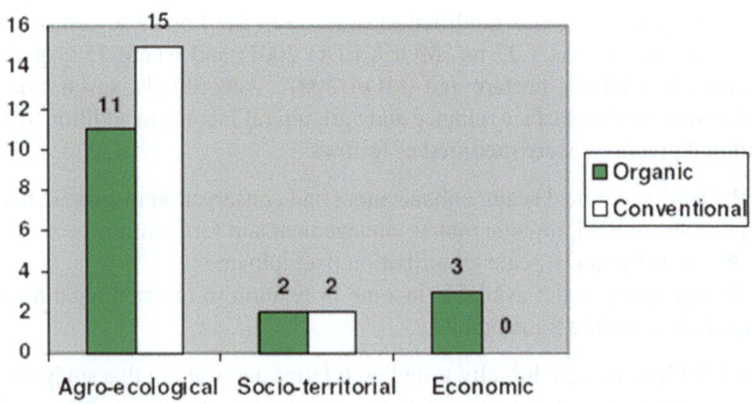

Fig. 17.1 Limiting sustainability value contributors to the sustainability scores for organic and conventional farms

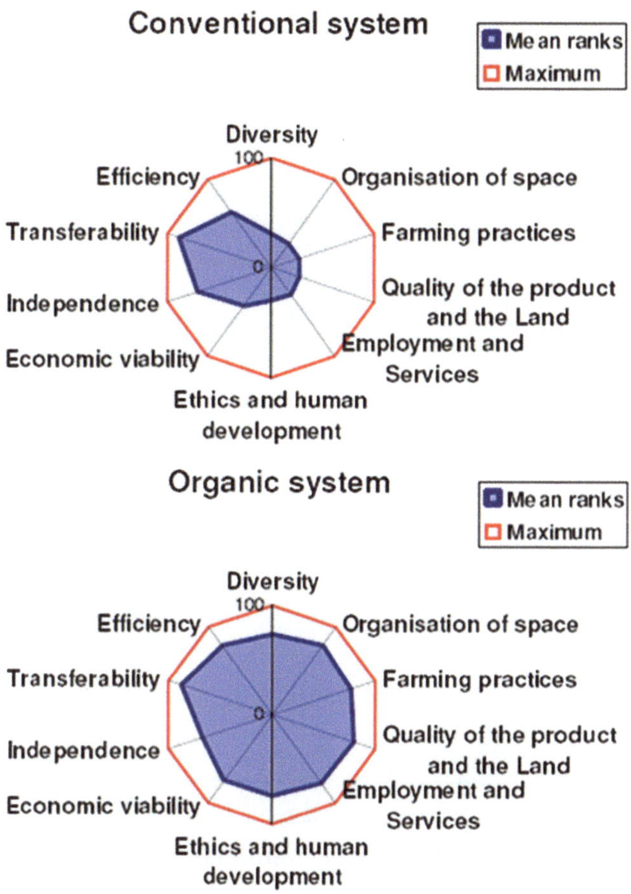

Fig. 17.2 Mean rank polygons of the studied components for the 17 tested farms under organic and conventional systems

organisation of space) and socio-territorial (quality of land and products, employment and services and ethics and human development) scales contributed to the better sustainability of the organic system versus the conventional one. No significant difference was revealed under the economic scales (economic viability, independence, transferability, efficiency).

17.3.1.3 Production System

The modified IDEA model revealed high variability in sustainability scores within a production system in agreement with Viaux (2003) who recorded the sensitivity of the IDEA model in a whole population of tested farms.

When comparing farms that scored the highest and lowest sustainability score, IDEA enabled us to identify the major differences or activities that contributed to

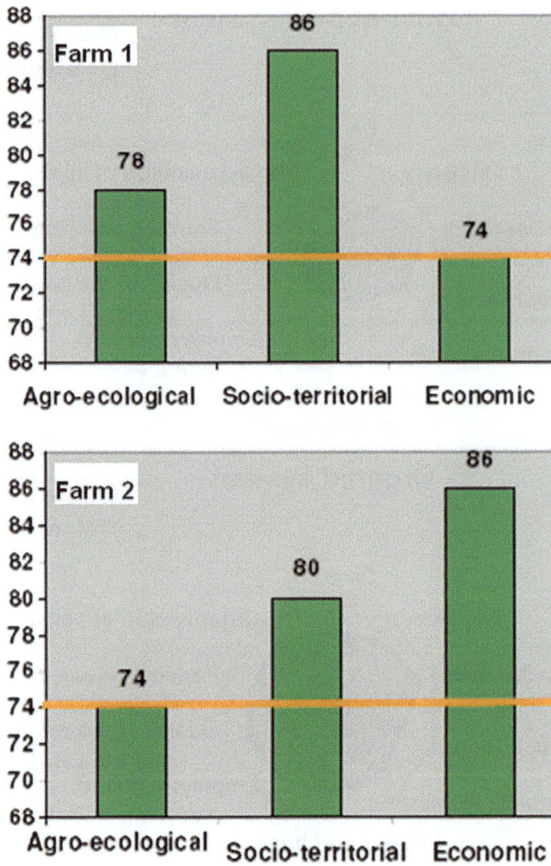

Fig. 17.3 Sustainability dimensions (agro-ecological, socio-territorial and economic) of two different organic farms

the variation of sustainability such as crop and animal diversity, energy and fertilizer dependence and others.

17.3.1.4 Sustainability Dimensions

The IDEA model exposed differences on the level of studied scales. Figure 17.3 represents the three sustainability dimensions of two studied organic farms. As can be seen, the sustainability scores can differ greatly according to the scale: in the first farm, the economic factor is the limiting one, while in the second farm it is the agro-ecological dimension. This comes in agreement to the findings of Zahm et al. (2006) about the practical difference between farms regarding studied dimensions.

17.3.1.5 Sustainability Components

It was shown that in some farms having the same sustainability scores related to the same dimension, the IDEA model revealed a sufficient sensitivity regarding the components of that dimension. This was also noted by Viaux (2003) reporting that no two farms resemble each other in the IDEA model.

17.3.2 Farmers' Self-assessment

The modified model proved to be user-friendly for farmers as a yearly self-assessment process. As a start, the farmers needed to be trained on the significance of each sustainability indicator and its calculation methods. In addition, regular recording of practices was essential for indicators' accuracy and especially the economic ones.

17.4 Conclusion

The modified IDEA model proved to be a useful assessment tool to guide farmers and development agents in assessing agriculture sustainability of small farms in semi-arid areas. It showed high sensitivity within the Lebanese context unveiling differences at the farm level between and within farming systems and identifying levels of intervention to improve sustainability. The model provided a simple but complete diagnosis that can serve as a decision support tool towards improving the sustainability of agricultural production systems in Lebanon.

References

Halberg, N., Verschuur, G., & Goodlass, G. (2005). Farm level environmental indicators; are they useful? An overview of green accounting systems for European farms. *Agriculture, Ecosystems & Environment, 105*, 195–212.

Hamadeh, S., Haidar, M., & Zurayk, R. (2007). *Research for development in the dry Arab region, The cactus flower*. Lebanon: Arab Scientific Publishers.

MOA (Lebanese Ministry of Agriculture). (2007). Agriculture in Lebanon: Yearly research report for Agriculture production 1998–2005. 122p. http://www.agriculture.gov.lb/arabic/enq_pro_98_05.pdf. Accessed 15 March 2007.

MOE, LEDO (Lebanon Ministry of Environment/Lebanon Environment & Development Observatory). (2001). Lebanon State of the Environment Report. http://www.moe.gov.lb/ledo/soer2001.html. Accessed 15 Mar 2007

Pacini, C., Wossink, A., Giesen, G., Vazzana, C., & Huirne, R. (2003). Evaluation of sustainability of organic, integrated and conventional farming systems: a farm and field-scale analysis. *Agriculture, Ecosystems & Environment, 95*, 273–288.

Salvatore, D. (2004). Globalization, comparative advantage, and Europe's double competitive squeeze. *Global Economy Journal 4*, (1), Article 2.

Viaux, P. (2003). Pour une agriculture durable. Vous avez dit durable, mais est vraiment mesurable? Arvalis. *Revue Perspectives Agricoles, 295*, 18–24.

Vilain, L., Girardin, P., Mouchet, C., Viaux, P., Zahm, F., Robert, F., Parent, P. (2003). La méthode IDEA Indicateur de durabilité des exploitations agricoles – guide d'utilisation – 2ème édition enrichie et élargie à l'arboriculture, à la viticulture, au maraîchage et à l'horticulture. (ed collectif O.), Educagri éditions, Dijon, 151.

Von Wiren-Lehr, S. (2001). Sustainability in agriculture – An evaluation of principal goal oriented concepts to close the gap between theory and practice. *Agriculture, Ecosystems & Environment, 84*, 115–129.

Zham, F., Viaux, P., Vilain, L., Girardin, P., Bouchet, C. (2004). The IDEA method (Farm Sustainability Indicators). A diagnostic assessment method to make the transition from the concept of sustainability to its assessment using indicators. PEER conference, Helsinki, Finland.

Zahm, F., Viaux, P., Girardin, P., Vilain, L., & Mouchet, C. (2006). *Farm sustainability indicators using the IDEA method.* Geneva: International Forum on assessing sustainability in Agriculture.

Chapter 18
Tendency of Production Decisions of the Farmers of the Southeast Pampa Region in Argentina Under Uncertainty Conditions

Mirna Mosciaro and Carlos Iorio

Abstract This chapter analyses the productive strategies of farms with intermediate levels of capitalisation to infer the resource allocation tendency. The analysis incorporates market and production risk considerations through application of two Minimization of Total Absolute Deviations (MOTAD) models. The first one includes conservationist recommendations on soil use; in the second one, these restrictions are not included. Comparative analysis of the representative productive system with the efficient plans reached places the farmer within a range of intermediate profit-risk levels solutions. While pastures and corn play an important role in the representative farm model, efficient plans tend to be less diversified, increasing acreage with wheat. Relaxation of the land use restriction allows obtaining any similar expected profit at a lower level of risk. However, solutions are more specialised than those with crop restriction. Results indicate that regardless of the producer's degree of aversion to risk soybeans and wheat will be the basis of productive plans. Specialised cash crop production will increase, unless efforts are made to promote adoption of sustainable practices by farmers.

Keywords Uncertainty • Decision • Sustainability • Farming system • Resources allocation

---

M. Mosciaro (✉)
Estación Experimental Balcarce, Insituto Nacional de Teconología Agropecuaria (INTA) Argentina, Ruta 226; km 73,5, Balcarce, 7620 Buenos Aires, Argentina
e-mail: mmosciaro@balcarce.inta.gov.ar

C. Iorio
Facultad de Ciencias Agrarias, Universidad Nacional de Mar del Plata Argentina, Ruta 226; km 73,5, Balcarce, 7620 Buenos Aires, Argentina
e-mail: ciorio@balcarce.inta.gov.ar

A.A. Marta-Costa and E. Silva (eds.), *Methods and Procedures for Building Sustainable Farming Systems: Application in the European Context*,
DOI 10.1007/978-94-007-5003-6_18, © Springer Science+Business Media Dordrecht 2013

18.1 Introduction

Medium- and large-scale farms in the southeast of the Pampa region in Argentina have been characterised by their diversified production. Integrating cropping and livestock activities via rotations of grain crops with pastures has been traditionally viewed as a strategy to conserve soil, maintain productivity and stabilise farm incomes.

However, since the beginning of this decade, there has been a progressive increase in area planted with annual crops. This expansion of cultivated land was mainly because of higher relative grains prices and improved soybeans and wheat yields during the 1990s. Technology for the production of soybeans, either single year harvest or like double-crop planted after wheat is harvested, allowed higher and more stable yields in the southeastern region of the Pampas. The tendency towards annual crops is reinforced by the introduction of French wheat genetics (called baguette varieties) that allowed increasing yields substantially.

The increase in area planted with crops has accelerated in recent years. Soybeans is leading this change by its advantageous cost/benefit ratio (Iorio and Mosciaro 2008), displacing other crops like corn or sunflower. The regional area sown with soybeans increased from 115,000 to 427,000 ha (+271%) between 2002 and 2007. At the same time, the corn-planted area decreased from 129,400 to 84,100 ha (−54%).

The recent increase of the soybeans area raises questions about whether intermediate farms will be able to maintain a diversified production plan.

The trend towards specialisation may have negative implications on the sustainability of farming systems. Agriculture practices are considered sustainable if they tend to maintain or increase soil organic matter levels over time (Robinson et al. 1994). The results of different rotations carried out in the region (Studdert and Echeverría 2000; Studdert 2003) reveal that sequences with high shares of soybeans have higher loss rate of organic matter due to biochemical characteristics and low carbon replenishment of its stubble. These authors conclude that negative effects of soybeans expansion can be reduced by diversifying the rotation with annual and perennial crops.

From the economic standpoint, diversification is a tool to mitigate the effect market and climate variability on farm incomes. Nevertheless, this strategy involves an implicit cost, which constitutes a suboptimal outcome (Anderson et al. 1977; Hardaker et al. 1997). This background makes necessary to analyse whether the fall on profit carries on a significant risk reduction.

The existing literature allows us to think that producers would choose diversified production plans, not necessary following conservationist criteria but because diversification constitutes an efficient strategy to reduce production and market risk.

The objective of this work is to analyse the trade-off between returns and risk for plans with different degrees of diversification and to infer the tendency of farms with intermediate levels of capitalisation in terms of soil use. We address this question using the classical approach to decision-making under uncertainty by modelling representative farm using a linear programming Minimization of Total Absolute Deviations (MOTAD) model.

18.2 Method

A whole-farm model of a typical farm of the southeast of the Pampa region is developed. The necessary parameters that define the representative farm were defined by a focus group formed by local producers and agronomic consultants (experts).

18.2.1 Analysis Model

The economic analysis is conducted using two different formulations of a linear programming model. The first formulation models the use of soil conservation practices, while the second does not include such practices.

The risk assessment is done through a MOTAD (Hazell 1971). MOTAD is used to analyse the trade-off between returns and risk of production plans with different degrees of diversification and to evaluate if farms that diversify more are more efficient in reducing risk than specialised farms. In order to validate our model, production plans obtained from the MOTAD are compared with the representative farm plan.

The optimisation done by the MOTAD works as follows: the absolute media deviation of returns (A) is taken as an indicator of benefits variability and interpreted as a measure of risk. Those plans that minimise A for given levels of expected return (E) constitute the efficient set of portfolios E–A. These portfolios yield the specified expect total margin (ETMs) assuming the minimum possible risks. This efficient set is further restricted by imposing a lower limit on the expected floor of the return (L), where L = ETM – 2 s and s is the standard deviation which makes that the return is 95% unlikely to fall below this floor (Baumol 1963).

Market and production are considered as sources of risk. Market risk is created by the variability of product prices, fertilisers and herbicides prices and by the variability of land-leasing fees. Production risks are created by the variability of crop and pasture productivity due to changing weather conditions. The impact of weather changes on pasture productivity, and in forage supply, is simulated through varying weight scale of the marketed animals. Random behaviour of these variables is emulated through stochastic Monte Carlo simulations. A total of 100 iterations were used given that these iterations assured appropriate levels of convergence.

18.2.2 Representative Farming System

The representative farm operates 700 ha, 500 ha are owned by the producer and 200 ha are rented. Only about 10% of the owned land is unsuitable for annual crops. The land allocated to each activity is as reported in Table 18.1.

Wheat and soybeans are planted through custom farming using direct seeding machinery. The remaining crops are planted using the farmer's own conventional

Table 18.1 Land utilisation (ha)

	Own land	Rent land
Total land	500	200
Cropping activities	400	200
Wheat	140	100
French varieties (baguette)		140
Traditional varieties		100
Corn	20	–
Sunflower	60	–
Soybeans	180	100
Double-cropping soybeans[a]	100	60
Livestock activities (effective use)		
Perennial pastures on land suitable for annual crops	50	–
Perennial pastures on land unsuitable for annual crops	50	–
Annual pastures (oat)	40	–
Stubble	60	–

[a]Soybeans planted after wheat is harvested

Table 18.2 Cropping activities: main inputs

		Wheat (baguette)	Wheat (traditional)	Corn	Soybeans	Double-cropping soybeans	Sunflower
Seed	(kg/ha)	150	150	20	90	110	4.5
Glyphosate	(l/ha)	3	3	–	7.5	5	–
Phosphate diammonium	(kg/ha)	100	100	80	50	–	50
Urea	(kg/ha)	180	140	120	–	–	–

tillage machinery. Table 18.2 shows the modal rate for inputs used by the representative farm. These rates were used in the specification of the MOTAD model.

Cattle production includes breeding beef and the fattening of steers and heifers up to the slaughter weight. The performance measures used to model the beef breeding herd and the feeder cattle herd are shown in Table 18.3. The fattening period of females and half of the males is 9–10 months, while the rest of males are feed from pastures for about 1 year and receive a supplementation of 4 kg per head per day of wet-corn kernel silage during the last 2 months.

18.2.3 Model Formulation

The MOTAD model is specified in a linear programming matrix form. Data and technical coefficients agreed as typical by the panel of local experts. In the case of harvest crop alternatives, two tillage technologies are considered for wheat: conventional

Table 18.3 Livestock: technical coefficients

Cows (head)	100	
Replacement heifers rate	20% – own	
Bull rate	4%	
Weaning rate	84%	
Weight gain (kg/day)		
Heifers	Variable, mode: 0.500	
Steers	Variable, mode: 0.550	
Sales own production	Weight (kg/head)	Sale month
	Mode: 310	December/January
Heifers	Mode: 340	December/January
Steers	Mode: 380	March/May

Table 18.4 Grain crops yields: frequency distribution

Wheat					Corn		Sunflower		Soybeans		Double-cropped soybeans	
	Traditional		Baguette									
	Conv	DS	Conv	DS								
Prob	ton/ha	ton/ha	ton/ha	ton/ha	Prob	ton/ha	Prob	ton/ha	Prob	ton/ha	Prob	ton/ha
10%	2.8	3.0	2.8	3.0	15%	2.0	15%	1.3	5%	1.4	10%	2.0
15%	3.3	3.5	3.8	4.0	20%	4.5	20%	1.8	15%	2.0	25%	6.0
50%	4.4	4.5	5.4	5.5	35%	6.5	40%	2.4	45%	2.8	40%	1.2
25%	5.2	5.2	6.5	6.5	20%	8.0	15%	2.8	20%	3.2	15%	1.5
–	–	–	–	–	10%	10.0	10%	3.0	10%	3.8	10%	2.0

Prob Probability, *Conv* conventional tillage, *DS* direct seeding

tillage and no-till seeding (direct seeding). Yield frequency distributions for each grain crop activity according to experts' opinions are shown in Table 18.4. Stochastic simulation takes into account the historic yield correlation between crop yields.

In addition to the described cattle activities defined in the model, sale of weaned calves (170 kg/head for females and 180 kg/head for males) and a short fattening period for heifers with spring supplementation (to be sold in October weighting 280 kg/head) are included. For cow replacement, two alternatives are considered: 15-month or 27-month heifers both produced internally.

Forage supply is modelled through independent activities according to soil quality requirements and seasonal production of each type of pasture crops. Table 18.5 shows the effects of variability in pasture productivity on weight sale of the marketed animals.

Simulation of output and input prices is based on triangular probability distributions. The most likely (mode), minimum and maximum values are set based on the typical sale (purchase) months for each output (input) between 1992 and 2009 (Table 18.6). Distribution parameters for input purchases are based on the typical

Table 18.5 Sale weights: minimum, most likely and maximum values (kg/head)

	Steers		Heifers		Light cow	Fattened cow
Sale months	January	March–April	January	October	March–Sept.	June–November
Minimum	290	320	270	250	360	400
Most likely	340	380	310	280	380	420
Maximum	370	410	330	300	390	450

Table 18.6 Outputs prices: minimum, most likely and maximum values

		Sale months		Minimum	Most likely	Maximum
Wheat		January–February–March	$/ton	315	470	758
Corn		May–June–July	$/ton	269	375	566
Sunflower		April–May–June	$/ton	458	747	1,145
Soybeans		April–May–June	$/ton	481	745	1,007
Calf	Female	February–March	$/kg	2.38	3.11	3.94
	Male	February–March	$/kg	2.75	3.38	4.16
Steer		January	$/kg	2.5	2.92	3.35
		April	$/kg	2.52	2.96	3.76
Heifer		January	$/kg	2.5	2.93	3.38
		October	$/kg	2.51	2.95	3.33
Cull cow	Light	March	$/kg	0.96	1.38	1.98
		September	$/kg	1.17	1.57	1.98
	Heavy	June	$/kg	1.57	1.98	2.87
		November	$/kg	1.59	2.11	2.72

purchase month between 2001 and 2009 (Table 18.7). All prices are expressed as April 2009 pesos (Wholesale Domestic Prices Index, basis 1993 = 100[1]).

Prices are simulated considering correlation between them. Sale prices of calves, heifers and steers showed a high positive correlation (higher than 75%) although their relation with the different cow categories was lower. In grains, a high correlation between wheat and corn (80%) was found.

Land rent is considered as a discrete variable tied to different soybeans price ranges because of the high correlation between these two variables. Wheat-soybeans double-cropping leasing fees are also simulated according to soybeans prices, but double-cropping land-renting fee is set as 50 US$/ha above the renting fee for soybeans production. The remaining cost items are considered constant and valued according to 2008 price average.

Both matrix specifications consider the existence of physical limitations to the expansion of production activities (maximum area according to availability, soil occupation times and soil aptitude). Matrix specifications differ in the inclusion of agronomic restrictions of maximum area for summer crops (225 ha) and for winter crops (225 ha). These restrictions aim at reducing soil degradation.

[1] Instituto Nacional de Estadísticas y Censos – Argentina.

Table 18.7 Inputs prices: minimum, most likely and maximum value

Prices	Phosphate diammonium ($/ton)	Urea ($/ton)	Glyphosate ($/l)
Minimum	1475.8	1122.0	14.0
Most likely	4195.5	2703.4	22.0
Maximum	950.8	609.0	9.0

Table 18.8 E-A efficient plans agronomic restricted MOTAD model: agricultural land utilisation

			ETM reduction (%)											
		Optimum	2.5	5.0	7.5	10.0	12.5	15.0	17.5	20.0	22.5	25.0	30.0
ETM (1,000 $)		774	755	736	716	697	678	658	639	620	600	581	542
A: risk measure (1,000$)		306	266	233	214	201	189	176	163	151	138	125	102
Rent land (ha)													
Single-crop		–	106	250	250	250	221	178	136	93	51	8	–
Double-crop		250	144	–	–	–	–	–	–	–	–	–	–
Activities on rent land (ha)													
Wheat	Bt SD	–	–	69	92	–	–	–	–	–	–	–	–
	Bt conv	250	196	85									
	Trad conv	–	–	–	63	141	117	95	73	50	28	6	–
Soybeans	Single-crop	–	54	96	94	109	104	83	63	43	23	2	–
	Double-crop	250	144	–	–	–	–	–	–	–	–	–	–
Activities on suitable annual crop own land (ha)													
Wheat	Bt SD	–	–	–	–	75	70	67	64	61	57	55	21
	Bt conv	225	225	225	185	127	121	113	105	96	88	80	59
	Trad conv	–	–	–	40	23	33	45	56	68	80	90	115
Sunflower		–	–	–	28	50	52	47	42	37	32	27	52
Soybeans	Single-crop	225	225	225	197	167	159	163	167	172	176	180	165
	Double-crop	213	134	132	131	118	101	97	94	90	86	83	1
Perennial pasture		–	–	–	–	8	14	15	16	16	17	17	38

A absolute media deviation of total margin, *Trad* traditional, *Bt* baguette, *Conv* conventional tillage, *DS* direct seeding

18.3 Results Analysis

Production plans that consider conservationist land use restrictions and conform the efficient E–A set are presented in Table 18.8. Production plans tend to become more diversified as risk and expected total margin (ETM) decreases.

Reaching the maximum ETM implies expanding baguette wheat-soybeans double-crop to the maximum allowed limit (whether owned or rented land). Nevertheless, rented land for double-cropped is included in the solutions only at highest levels of benefit but also at the highest levels of risk. On the other hand, soybeans single-crop has relatively low variability and high-expected unitary margin (EM) which makes it the unique summer crop even in those solutions with the lower risk levels.

Table 18.9 E–A efficient plans agronomic unrestricted MOTAD model: agricultural land utilisation

			ETM reduction (%)										
		Optimum	2.5	5.0	7.5	8,9	10.0	12.5	15.0	17.5	20.0	25.0	30.0
ETM (1,000 $)		850	828	807	786	774	765	743	722	701	680	637	595
A: risk measure (1,000$)		405	356	313	277	261	251	235	219	203	189	162	134
Rent land (ha)													
Single-crop		–	103	232	250	250	250	250	250	230	197	127	40
Double-crop		250	147	18	–	–	–	–	–	–	–	–	–
Activities on rent land (ha)													
Wheat	Bt conv	250	147	59	–	–	–	–	–	–	–	–	–
Soybeans	Single-crop		103	191	250	250	250	250	250	230	197	127	40
	Double-crop	250	147	18	–	–	–	–	–	–	–	–	–
Activities on suitable annual crop Own land (ha)													
Wheat	Bt SD	–	–	–	8	70	120	95	85	75	64	53	49
	Bt conv	450	450	450	414	328	240	194	178	140	132	110	85
	Trad conv	–	–	–	–	–	31	103	125	130	129	134	116
Sunflower		–	–	–	–	–	–	4	20	20	20	29	26
Soybeans	Single-crop	–	–	–	27	52	60	54	41	62	75	99	153
	Double-crop	438	438	438	358	303	285	235	163	147	135	113	93
Perennial pasture		–	–	–	–	–	–	–	–	23	29	26	21

A absolute media deviation of total margin, *Trad* traditional, *Bt* baguette, *Conv* conventional tillage, *DS* direct seeding

In the farmer's own land, single-crop soybeans and baguette wheat are allocated the majority of cropping area in every efficient plan. Nevertheless, with an ETM reduction equal to or larger than 7% plans become more diversified including sunflower and pasture. As risk and total benefit decrease, crop combinations maintain a similar proportion between regular soybeans and sunflower, while double-cropped soybeans reduces progressively its participation.

Pastures are included in efficient E–A solutions when the maximum ETM reduces by 10 or more. However, pastures occupy significant amounts of farmland only in plans with low risk levels. Together with the incorporation of pastures, feeder cattle are also included on high quality land. Plans with ETM reductions of 30% or more increase and diversify fattening activities.

Predictably, when the model that is free of crop rotation constraints achieves the maximum ETM, the solution allocates all suitable land to the activity with the highest expected margin: baguette wheat-soybeans double-crop (Table 18.9).

This double-crop occupies more than 85% of the owned land suitable for annual crops up to a 15% reduction of maximum ETM. For further ETM reductions, the model allocates more land to regular soybeans. In these efficient plans, the area of pastures increases slightly with respect to those with land use restriction.

Like in plans with agronomic restriction, the double-crop in rented land is present only at high-risk levels. The payment of a premium for the longer period of land use makes double-crop more risky than single-crop soybeans. In the own land, double-crop uses an important proportion of area in most efficient E–A plans.

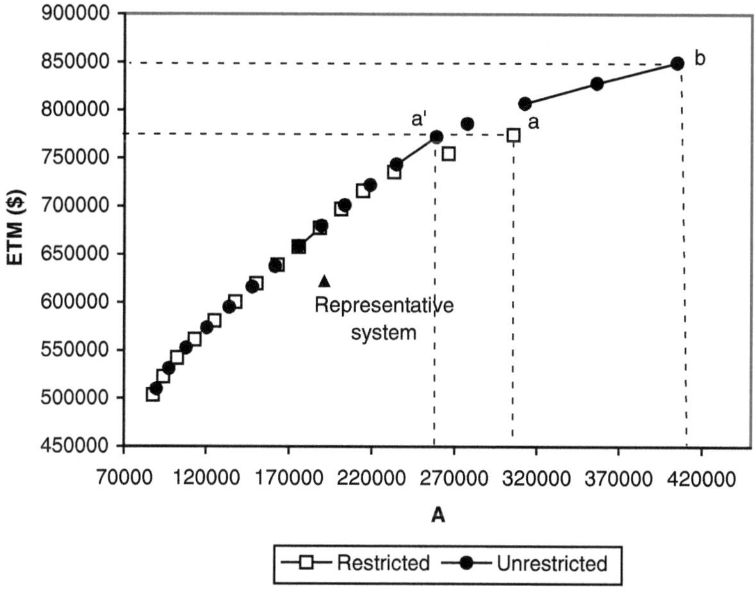

Fig. 18.1 An efficient set of plans for agronomic restricted and unrestricted MOTAD model

Figure 18.1 shows that in absence of land use restriction, the maximum ETM (point b) is 9% higher than the restricted maximum ETM (point a). However, reaching the unrestricted maximum involves a more than proportional increase of 25% in the absolute media deviation (A). However, if conservationist restrictions are ignored (i.e. unrestricted model specification), it is possible to reach an ETM equal to the maximum restricted ETM, but reducing the risk level by 15% (point a'). In the production plan corresponding to point a, single-crop soybeans in the own land is replaced by baguette wheat using a higher proportion of its stubble to increase cattle rearing, the lower risk level activity.

18.3.1 Comparisons of Efficient A–E Solutions with the Representative System

The productive plan of the modal farm agreed by the panel of experts is shown in Fig. 18.1 according to its expected total margin and absolute media deviation. The productive plan followed by this representative farm yields a lower ETM and is riskier than the plans included in the efficient set. The modal farm could benefit by reducing its level of risk by 21% while maintaining the same ETM ($620,600) or by increasing its ETM by 10% while maintaining the same risk exposure.

However, the combination of activities of the representative farm is similar to restricted plan yielding $619.500 (Table 18.8). There are two notorious differences between efficient and representative plan. First, while the representative plan grows the wheat-soybeans double-crop on rented land, the efficient plan uses the rented to produce single-crop soybeans. High gross margin variability of soybeans in the double-crop explains the higher risk taken by the representative farmer. The second difference is that the efficient plan tends to be more specialised in crop activities increasing wheat area, while the representative plan allocates more land to pastures and cattle fattening activities.

On the other hand, efficient plans tend to be more specialised in crop activities increasing wheat surface, while pastures and cattle fattening activities get an important place in the representative farm plan.

18.4 Conclusions

The analysis of efficient return-risk solutions shows that relaxing land use restriction allows obtaining expected profits similar to those of the restricted maximum but with a lower level of risk. However, these solutions are more specialised than those with crop restriction even at intermediate or low-risk level.

This result is against the assumption that uncertainty in yields and/or prices leads to more diversified production plans. Strategies to reduce risk may not be necessary consistent with conservationist practices.

Comparative analysis of the representative production plan with the efficient plans places the farmer within a range of intermediate profit-risk levels solutions. Differences, found between representative farm and efficient plans, may be partially explained because the model does not consider some particular considerations, such as financial and labour restrictions. However, the most important cause may be that still farmers' decisions are motivated by soil conservation goals. It is also likely that farmers having medium- or large-scale farms consider the inclusion of pastures in rotation, despite the revenue decrease.

Results of this chapter provide additional elements to explain the observed tendency towards the specialisation in annual cash crops and suggest that such tendency will continue, unless substantial efforts are made to promote sustainable land management practices. These promotion efforts should focus at farmers with different production scales and degrees of risk aversion.

Acknowledgements This work benefited from the support of the *Agence Nationale de la Recherche* (France) under the SYSTERRA program with the reference ANR-09-STRA-04, the *Instituto Nacional de Tecnología Agropecuaria* (Argentina) AEES-302442 project and associated projects.

References

Anderson, J. R., Dillon, J. L., & Hardaker, j B. (1977). *Agricultural decision analysis*. Ames: Iowa State University Press.

Baumol, W. (1963). Expected gain-confidence limit criterion for portfolio selection. *Management Science, 10*(1), 174–182.

Hardaker, J. B., Huirne, R. B. M., & Anderson, J. R. (1997). *Coping with risk in agriculture*. Wallingford: Cab International.

Hazell, P. B. R. (1971). A linear alternative to quadratic and semivariance programming for farm planning under uncertainty. *American Journal of Agricultural Economics, 53*, 53–62.

Iorio, C., & Mosciaro, M. (2008). Aportes a la discusión sobre diversificación productiva o monocultura en el sudeste bonaerense desde una perspectiva económica. *Revista Argentina de Economía Agraria – Argentine Agricultural Economics Review. Nueva Serie, 10*(1), 139–163.

Robinson, C. A., Cruse, R. M., & Kohler, K. A. (1994). Soil management in sustainable agricultural systems. In J. L. Hatfiled & D. I. Karlen (Eds.), *Sustainable agriculture systemas* (pp. 109–134). Boca Ratón: Lewis Publishers.

Studdert, G. A. (2003) "Sojización" ¿Un riesgo para los suelos del S. E. bonaerense? In: Actas 20º Jornada de Actualización Profesional sobre Cosecha Gruesa (pp. 94–103), Mar del Plata, Argentina: EEA INTA Balcarce – Fac de Ciencias Agrarias UNMP.

Studdert, G. A., & Echeverría, H. E. (2000). Crop rotations and nitrogen fertilization to manage soil organic carbon dynamics. *Soil Science Society of America Journal, 64*, 1496–1503.

Part VII
Final Considerations

Chapter 19
Methodologies for Building Sustainable Farming Systems: The Main Critical Points and Questions

Ana Alexandra Marta-Costa and Emiliana Silva

Abstract Sustainability is a dynamic concept, seeking to achieve a balance in space and time, of the environmental, economic and social dimensions. In this context, farming systems are faced with a double (and often contradictory) challenge to be successful: socio-economic performance has to be maximised while environment and natural resources needs to be protected. When choosing the best alternative, the method of evaluation plays a key role. This chapter synthesises the approaches exposed in the book for building sustainable systems in three classes and identify the main critical points of them as a whole. The chapter ends also with the identification of the main questions derived of the universe of used approaches that can be used in future reflections and discussions regarding the studied theme.

Keywords Sustainability • Sustainable farming systems • Sustainability farming assessment • Methods • Approaches • Decision-making • Future

19.1 Critical Points for Building Sustainable Farming Systems

Sustainable development is currently the dominant paradigm that guides the development planning. The Bruntland Report criticises the model adopted by the developed countries and advocates a new type of development that can sustain progress across

---

A.A. Marta-Costa (✉)
Centre for Transdisciplinary Development Studies (CETRAD), University of Trás-os-Montes and Alto Douro (UTAD), Quinta de Prados, Apartado 1013, 5001-801, Vila Real, Portugal
e-mail: amarta@utad.pt

E. Silva
Centre of Applied Economics Studies of the Atlantic (CEEAplA),
University of Azores, Rua Capitão João d'Ávila,
9700-042 Angra do Heroísmo, Açores, Portugal
e-mail: emiliana@uac.pt

A.A. Marta-Costa and E. Silva (eds.), *Methods and Procedures for Building Sustainable Farming Systems: Application in the European Context*,
DOI 10.1007/978-94-007-5003-6_19,

the world. The concept of sustainable development or sustainability, which aims to meet the needs of the present without compromising the ability of future generations to meet their own needs, was established as official to correct the effects of the ecological crisis (World Commission on Environment and Development – WCED 1987).

However, it still persists several difficulties to give a concrete content to sustainability. Serious efforts allowed a progress in defining the concept, such as those by Altieri (1994) and Hansen (1996). Despite some controversy, the debate around sustainability brings awareness of the complexity and interaction of the different dimensions (environmental, economic and social) and a need for a more integrated action between them (Altieri 1994). Also, the European Council of Stockholm acknowledged that, in the long-term, environmental protection, economic growth and social cohesion must go hand in hand (Commission of the European Communities – CEC 2001). The practical realisation of this objective requires that economic growth supports social progress and respects the environment that social policy underpins economic performance and that environmental policy is cost-effective (CEC 2001). These seem to be the three basic dimensions – environmental, economic and social – regarded as the tripod of sustainability, yet used in the recent documents of the CEC (2009). In fact, sustainability is a dynamic concept, seeking to achieve a balance in space and time, of the environmental, economic and social dimensions. It is not a fixed concept but vulnerable and influenced by other factors.

In this context, farming systems are faced with a double (and often contradictory) challenge to be successful: socio-economic performance has to be maximised, while environment and natural resources need to be protected. These are two conditions to obey to more sustainable farming systems. When choosing the best alternative, the method of evaluation plays a key role.

This book demonstrates an impressive range of topics and approaches to build sustainable systems, especially regarding their efficiency and their results. They can be grouped in three classes. In the first group, it was used or mentioned indicator-based methods for the sustainability assessment of agricultural systems. Some of them are integrated in more rigorous and complex frameworks as the *Arbre de l'Exploitation Agricole Durable* (ARBRE), the *Diagnostic de Durabilité* (sustainability diagnosis) of the *Réseau de l'Agriculture Durable* (RAD), the Framework for the Evaluation of Sustainable Land Management (FESLM), the *Indicateur de Durabilité des Exploitacions Agricoles* (IDEA, sustainability indicator of farms), the Indicator of Sustainable Agricultural Practice (ISAP), the Multiscale Methodological Framework (MMF), the Response-Inducing Sustainability Evaluation (RISE), the Sustainability Assessment of the Farming and the Environment (SAFE) and the Sustainability Solution Space (SSP) for decision-making method. Other method identified in this group, using a combination of other techniques, was the DELTA method. All of these approaches include the economic, environmental and/or social dimensions of sustainability.

The second group of approaches is related to linear programming and multi-criteria decision-making (MCDM) methods. MCDM is a methodology aimed at supporting decision-makers faced with numerous and sometimes conflicting criteria that can be formally incorporated into the management planning process (Romero and Rehman 1989). There were used the multi-objective programming, multi-criteria decision method combined with life-cycle assessment (LCA), the Minimization of Total Absolute Deviations (MOTAD) and the mathematical programming approach. Also it was coupled an economic optimisation model (Opt'INRA) with a model assessing nonrenewable energy (NRE) consumption and greenhouse gas emissions (PLANETE).

Finally, the life-cycle assessment (LCA) of a product and the energy analysis and footprinting constitute the third group of structured tools used along this book.

All these procedures were used alone or with combinations between them in a structured way to complementary their actions and improve efficiency and/or effectiveness. Each essay enhances the appreciation of great international current efforts for building sustainable farming systems. The creation, use and reformulation of several approaches are significant attempts for the development of sustainability in agricultural sector. Besides, its building, application, comparison and assessment make the concept of sustainability useful, allowing to highlight the several dimensions and to identify problems and their potential solutions to conduct to more sustainable futures and to more sustainable methodologies.

Other final remarks can be detached in order to get future reflection and discussion:

- Emphasis on the environmental area – many available methodologies and procedures of global scale are majority based in environmental features. The economical and mainly the social aspects are not as well represented in the approaches mentioned in the chapters of this book.
- Validity of the procedure for measuring indicators – indicators are the main tools used for assessing sustainability. They can be used alone or integrated in indexes or structured methodologies. However their measurement process or sources of information allow relevant questions, jeopardising the obtained values and all used procedures.
- Subjective nature of the approach – some methodological alternatives evidence a strong subjective character. This is due to the flexibility of procedures used to select, to measure and/or to monitor indicators that can vary with the subject, the context of the evaluation and/or evolved stakeholders. This may become advantageous because it makes possible to adapt the methodology to the reality under study. On the other hand, subjectivity may emphasise one or another side, minimising other aspects also important for sustainability and removing the possibility of comparison of studied object at different levels.
- Systemic approach – some methodologies do not show concern about the systemic approach. They do not analyse the overall systems attitude, with emphasis on relationships and interactions that occur between its various components. Also, the interdisciplinarity is not yet present in all methodologies.
- Interaction between many alternative approaches – in most cases, there are no interactions between several methodologies when they are built. Knowledge and an integral view of them can generate methodological alternatives able to maximise its benefits and to minimise its drawbacks. This is a step needed to build better future for farming systems. However, many efforts to pursue this objective were also highlighted with some of the approaches applied in the book.

19.2 Building Sustainable Farming Systems: Final Questions

This book started with three main questions: (1) What methodologies are being developed for building sustainable farming systems? (2) What are their results? (3) Are there effective proposals to build sustainable farming systems efficiently? Consequently, current research was shown addressing the issues that focus on building, application, comparison and discussion of methodologies applied in the assessment of sustainability as a global concept (economic, social and environmental). In the last 18 chapters, it was seen that some of the used or developed approaches have advantages in particular aspects and the remaining methodologies in other features. The real contexts ask for special methods, and theoretical evaluations need global dimensions. However, the universe of approaches used for building sustainable farming systems exposed on this book allowed new main questions that can be used in future reflection and discussion about the studied theme:

1. Is the combination of sustainability assessment methods a way to an enhanced and more real sustainability assessment than the use of an isolated or individual methodology?
2. What procedure is more efficient for the sustainability assessment? The identification or building of just one sustainability assessment method common to general systems and activities or sustainability assessment methods specific to each system or activity?
3. Are there systems (e.g. the organic systems) or specific practices really more sustainable than the conventional systems or practices in all situations? As scientists should we give incentive to the farmers of organic systems?
4. How can decision-makers use the approaches for building sustainable farming systems?

In the current times, the final remarks about these questions are that it remains impossible, in absolute terms, to sustain which of the defined method(s) is(are) unquestionably the finest. Therefore, the one that in a given moment or circumstances seems to be the best would not be at a different moment in time and in other circumstances. This means that it cannot be concluded at all, what procedure or what methods or combinations should be elected in guiding sustainability assessment.

Also, the sustainable farming systems are yet unknown. There are systems more economical sustainable, but others are preferable to the environmental dimension. There is unidentified engagement of the evaluation areas of sustainability as the dynamic of sustainability concept is not yet well known. All methodologies and farming systems have advantages and disadvantages, within different socio-economic and environmental conjunctures and within special objectives. All of them are valuable and needed attempts to understand sustainability, to understand the development that we need for our future farming.

References

Altieri, M. A. (1994). Bases Agroecologicas para una Produccion Agrícola Sustentable. *Agricultura Técnica, 54*(4), 371–386.

CEC. (2001). Commission Regulation (EC) N.° 1837/2001 of 10 September 2001 amending Regulation (EEC) N.° 2237/77 on the form of farm return. *Official Journal of European Communities* L 255, 24.9.2001.

CEC. (2009). *Mainstreaming sustainable development into EU policies: 2009 Review of the European Union Strategy for Sustainable Development.* Communication from the Commission to the European Parliament, the Council, the European Economic and Social Committee and the Committee of the Regions, COM/2009/0400 final.

Hansen, J. W. (1996). Is agricultural sustainability a useful concept? *Agricultural Systems, 50*, 117–143.

Romero, C., & Rehman, T. (1989). *Multiple criteria analysis for agricultural decisions.* Developments in Agricultural Economics 5. Amsterdam: Elsevier.

WCED. (1987). *Our common future*. Oxford: Oxford University Press.

MIX
Papier aus verantwortungsvollen Quellen
Paper from responsible sources
FSC® C105338

If you have any concerns about our products,
you can contact us on
ProductSafety@springernature.com

In case Publisher is established outside the EU,
the EU authorized representative is:
Springer Nature Customer Service Center GmbH
Europaplatz 3, 69115 Heidelberg, Germany

Printed by Libri Plureos GmbH
in Hamburg, Germany